FINDING THE NEXT STEP IN OUR EVOLUTION

By
Jeff Cooke

PREFACE

Our species is currently evolving whether we can acknowledge this yet or not. My interpretation of this phenomenon is only one out of eight billion human experiences. Our current growth in the evolution of consciousness may seem like a complete mess on the surface of human life at this time. Yet perhaps this is a part of natural selection creating for itself a means, or a need to evolve beyond the current level of human consciousness. The interpretations in this book are simply one grain of sand on a giant beach of human interpretations within this subject of consciousness. We are evolving whether I interpret this subject or not. I simply felt compelled to attempt to write about this miraculous moment in our species' existence. In hope that this might help and inspire some who are ready to look more deeply within themselves.

TABLE OF CONTENT

CHICKEN & EGG

"The important thing is to not stop questioning.

Curiosity has its own reason for existing."

-Albert Einstein

When the sound of a cricket just outside of our home late at night is resisted through thoughts that insist it is disturbing and shouldn't be there. It can become impossible to fall asleep. If, however, the sound is accepted for what it is, it will no longer have this power over us. The sound may even become relaxing and harmonious, making it easier to fall asleep. When the mind is too restless, nothing can be harmonious. When the mind can become still enough, accepting enough, the entire world has the potential of becoming harmonious, regardless of the current circumstances.

The scientific answer to the question, "Which came first, the chicken or the egg?" is still debated. Yet many scientists claim that it is neither, because a bird that was evolving from an earlier species into the chicken species, laid one of the first chicken eggs. This was a process that occurred not just from one occasion, but hundreds of millions to maybe billions of eggs for the entire species of proto-chickens to slowly evolve into the chicken species we know today.

The amount of time this took was just a blink of the eye in comparison to the age of the Earth or the age of the Universe. Yet this transition must have lasted eons from a single bird's perspective. Not that any individual bird would know that their species was evolving. However, on some deep level, their DNA was changing. Not just by chance, but through some form of deeper intelligence than the almost chickens would likely understand on the surface levels of their lives.

This question of which came first, was at one point asked in order to try and momentarily stop someone's thought process with an unanswerable question. Not used very often, but if asked at the right moment, had the potential to create a short gap between thoughts, and allow the possibility for something new to come in. Like the question, "If a tree falls in the forest, and no one is around to hear it, does it make a sound?" Today we can place recording devices in the forest and record whatever sounds we wish. Before modern science however, these questions were considered unanswerable. Often used with the understanding that an overactive mind sees anything and everything as fuel for more non-stop thinking. Presenting an unanswerable question could introduce a moment of peace from this endless noise.

We could say the mind loves to think, or more accurately, the mind is obsessed with its own thinking, in many cases, unable to stop. Even if, and in some cases especially if it involves creating a conceptual problem that doesn't exist. With this understanding, the more we can become aware of our thinking patterns, the less we will find ourselves lost within problematic thinking. One key ingredient is to also become more aware of the gaps between thoughts, which can allow moments of peace to come into our daily lives. Zen Masters, as well as other Spiritual teachers throughout history, have used questions like these in order to help their disciples realize the importance of noticing the peace or silence that is within these gaps between thinking.

However important any potential solutions might be towards a current situation, it never seems to be quite as important as the mind's need to compulsively think about it. Often holding a certain opinion as a kind of trophy to cling to, and identify ourselves with. Defining what is seen as right and wrong within a situation, taking away any need to ask new questions, making it more difficult to learn any new information that the mind doesn't want to learn. Once the mind identifies with an opinion, it can become impossible to let it go, until we can see these behaviors more clearly within ourselves. Repetitively reestablishing conceptual beliefs with repetitive overthinking becomes more important than solving a situation in many cases. It becomes "my way or the highway" as the saying goes. This leads to a lot of dysfunctional behaviors within society. Any situation that has one side against another could become resolved rather quickly and peacefully if most humans were not lost in these mental behaviors of identifying with thought. Awareness of this is the key to ending these

3

sorts of dysfunctional behaviors within our lives and our world. However, it is only on an individual level that we move beyond these patterns. No one can show another person what they are not ready to see.

Like the chicken and the egg, we are all individually at different positions along a collective evolution that has had its seeds planted ions ago and is reaching the threshold of dysfunctionality and chaos that will allow our species to take the next step forwards in the evolution of consciousness on planet Earth. Through becoming more aware of ourselves, more aware of the mental dysfunctionalities most humans are lost within. Starting from the time that the earliest hominins began living with complex minds. A new challenge was created, an ability to have these minds without losing ourselves within them. Since that point there have been many practices of spiritual wisdom throughout history and prehistory that point in this direction, some still existing today. However, no amount of wisdom is useful until we are fully ready to understand it.

Some of us are more ready than others, and it is simply not possible for anyone to be more ready than they currently are in order to see their addictions with overthinking, to be able to see how they are not the thoughts that their mind is having, but the deeper awareness that is life itself. Those who are ready are beginning to evolve beyond overthinking addictions, finding that most of life is much more peaceful than many thoughts would have them believe. When we are no longer unknowingly dragged into negative thinking, the world becomes much more beautiful as it is, and the need to always have a conceptual story or answer to everything falls away more and more of the time. We do not lose the ability to think, we simply lose the obsession with always needing to think at every moment of the day. Gaps between thinking allows our quality of thoughts to improve.

The current stage of human evolution that we have been in for as long as we've had a human mind could be called the "Ego Stage". The word ego is mostly used today to point out when someone is being narcissistic, acting rudely, or is using their "lizard brain" (the least evolved quartext of the mind). Sometimes simply when someone is making a situation all about themselves. These all point in the direction of what ego is to some degree. Unfortunately, what most people see as ego is a crude definition that does not point anywhere near the deeper reality of what the ego actually is. Yet it makes perfect sense that most humans, while within the ego stage, would be oblivious to its deeper definition. The ego is a personalized identification with every thought the mind thinks. Nothing more, nothing less. The belief that the identity we hold with our thinking patterns is who we are, and all that we are. We may wear many different costumes throughout our lives, yet identification with thoughts is the most impactful. It is this mental costume that tells us who we are, while keeping us from truly knowing ourselves.

Where does a thought originate? About 90 to 99 percent of the thoughts we think do not come from our own mind but from somewhere or someone else. Yet as we repeatedly think these thoughts, we can't help but put an added identity onto the ones our mind agrees with, and some sort of label that is "not me" on the thoughts the mind does not agree with, or does not identify with. This will always add the need to defend a stream of thought patterns as if conceptual ideas are who we actually are, which has in turn, made our thoughts and actions more dysfunctional and chaotic in general as we regularly must repeatedly defend an illusory identity of ourselves. The unconscious, or not yet acknowledged, need to identify with a conceptually filtered understanding of our lives, as well as the world in general, can only lead to various degrees of feeling incomplete in many ways.

Until this illusory sense of self is fully seen for what it is, we are unable to understand why life always feels like something is not quite right. Why something always seems to be missing in life. The mind will try to fill this apparent void with whatever it can think of. Yet every addition, subtraction, or change to life for this sake will only be satisfying for a short time before thoughts and feelings of incompletion come back in, creating yet again the same deep feelings of not being whole. Until a deeper understanding of ourselves is ready to be realized, life can appear to be meaningless for much of the time from the ego's point of view. Many of our thoughts, whether noticed or not, can easily gravitate towards negativity. Thoughts like, "This didn't happen the way it should have", "I didn't get what I should have", "I or they shouldn't have acted that way", to only mention a few. Ego will very willingly cling to a negative perspective and a negative identity to some degree. As long as there is a mental identity for it to cling to, a negative one will work just fine. Unfortunately, in many cases a negative identity is easier to cling to than a positive one. The conceptual "I" is more easily defined when the story is that the world is against "me". Not that optimism cannot be egoic as well, just that the mind is very dualistic, and until noticed, this phenomenon swings back and forth at some level for every ego.

A mind is a wonderful thing. There are many reasons to learn new things and be productive in many different intellectual ways. However, a mind that we fully identify with can't help but create chaos for itself and others to some degree on a regular basis. Overactive thinking has also cut us off from a key ingredient in life. The deeply peaceful feeling of life itself flowing throughout this body we inhabit which is the source for a deep level of inner peace that the conceptual mind cannot understand. Because it is not something it can grasp conceptually, it is not a thing, nor is it something it can do. It is not the restlessness that the mind often identifies itself with, it is simply Being. Simply becoming aware of our own inner stillness that is always within. For most humans, this is always covered up by overthinking. Words are very limited in describing this but they can be helpful for pointing toward this inner awareness. As you read this book, remember that the words themselves are not what's being pointed towards. They are simply here as an aid to help you see what is more deeply within. In many cases, it is the space between the words that are most helpful. This should continue to make more sense as you continue reading.

Inner stillness, one label where the label is not quite what it is, but a conceptual pointer towards what it is, has the ability to calm the restlessness of the mind whenever we are able to become still enough within ourselves to notice this aliveness within the body. Something the mind tries to avoid because there is little need for endless thinking about the past and future when we can be content with the inner peace we can find within ourselves now. Thinking is all that the mind knows. To the mind, it can feel like it will lose itself if it stops thinking. In truth, nothing is lost except for the restlessness of the ego. If you take a moment now, and place your attention on your breathing. You may begin to feel this deep peace that these words can point towards, but can only be felt beyond the conceptual idea of.

Simply hold your attention on the lungs filling and emptying of air. As your attention shifts more into the body from the mind, a deeply peaceful feeling begins to emerge from a deeper place than thought or emotion. This is the stillness that is more deeply who you are than any thought or emotion that is experienced. It does not replace thoughts and emotions, you are simply noticing something deeper that has always been here, simply covered up by endless thoughts. The more we anchor ourselves within this realm of our Being, the less we are lost in conceptual confusion of who we think we are, and the more peaceful life naturally becomes. Not without challenges, but with less resistance-based reactions towards the challenges in life.

Are you addicted to your thinking? Can you notice when a single thought turns into many more? Are you able to notice how repetitive your thoughts can be? Can you notice how certain thoughts always bring with them the same emotional reactions and how certain emotions always seem to spark the same thought patterns? This can be a lot like the chicken and the egg. It can be impossible to know which comes first. More importantly, do you know just where your mental and emotional suffering begins? The sort of heaviness that many thoughts and emotions can seem to be fueled by.

Our mental and emotional suffering does not come from the challenges or physical pain we may encounter in life, but from our resistance towards what the mind doesn't want to face. No life is without challenges, or painful experiences to some degree. That being said, how challenging is it for most people to see the difference between a situation, and their mind's habitual resistance patterns reacting towards a situation? This awareness is the key to our next step in evolution. The ability to be conscious enough to have a mind, without being lost within its reactive based thinking patterns. A

9

certain amount of alertness is required to not be pulled along by every stress-filled thought, a certain amount of awareness. This can require practice, but once we become ready, we become willing. The mind will torture itself enough that we begin to shift beyond the identity simply through a need to not be lost within this repetitive negativity, which begins to give itself away more and more as we practice.

Overthinking, as well as negative thinking patterns within our species, is not a curse but a natural part of evolution in order to move from one level of consciousness to the next. Comparable towards what earlier hominins must have experienced as they started cooking their food, and brain size was able to increase. Where we are today, is the next step in this process. Our brains might be about as big as they need to be, but more important is how lost within our thinking we have become without even realizing it. It can be very difficult to know when we are within a dream, until we begin to awaken. Conceptually based words are very limited in describing a conceptually based dream. However, most humans today are lost in this so-called dream of overthinking. So much so, that 90 to 99 percent of our personal and societal problems arise, not because of the situations we encounter, but how the habitual mind patterns that we identify with as who we are, react towards these situations.

Try not to read this the wrong way, we can solve many problems through thinking, but when at the root of the problem is identification with thought, that causes most of the dysfunction within ourselves as well as society. We will need a new level of consciousness in order to create a world no longer lost within perpetual problem-making. This is achievable, but not by convincing others how to act or what to do. This change requires enough individuals to become aware enough of themselves for society in general to move beyond our current societal patterns of ego. Every individual who becomes more aware is bringing the collective closer towards this fundamental change.

Growing up, I was convinced a lot of the time, and in many different situations, that I was doing something wrong, or never quite getting things as right as I should be. I didn't have a particularly tough or challenging childhood, but did develop depression like many teenagers do. I knew something wasn't quite right, but I didn't really know what, or why. Essentially, the dream of ego was holding together the various degrees of suffering it was supposed to for its own purposes of survival, you might say. Ego wants to survive just like any other form of life. The normal way of living as most people understand, was expressing itself as it should. The process of increasing self loathing became stronger and stronger as I got older. It became more and more difficult to function in life without negativity overwhelming my thoughts and emotions.

It took until my late twenties, and help from others, to begin to realize just how extremely overactive and reactive our minds can actually be. How identifying with this creates confusion and resistance within ourselves. As well as how the mind interacts with emotions in an unconscious or un noticeable way. Many negative emotions become overlooked in a way, or not seen directly as they are, as they become normalized most of the time. Through the mind's need to impose an identity onto the emotions we experience within any given situation. Repetitive emotions easily become a confirmation for the identification we hold within our negative thought patterns, while negative thoughts continue to fuel negative emotions. Without a deep enough investigation within ourselves, these cycles occur as a neverending conformation of how situations, or how our life in general, shouldn't be the way it is. Negative or positive, any emotion the body expresses becomes grasped by the mind as a part of its definition of the mental "I", our illusory sense of self.

Most people think they are thinking, but really, thinking is happening to them, it is something we cannot simply stop. If we cannot stop it, we are not in control of when the mind is thinking. So, most of the time, thinking is happening to us. It has a certain momentum to it because of how long our species has been overthinking. It has long since been within our genes to overthink as much as we possibly can in order to give as much attention as possible to the ever-thinking mind. Repetitive thinking is the real addiction within humanity, the true gateway drug. The one that leads us to almost every other addiction in our lives because of how much we identify ourselves with most thoughts. The only thing a mind truly understands is what we could call the dimension of thought. Making life seem more conceptual than it actually is. More easily confusing concepts with reality itself. We could say that reality is always imprinted to some degree with our conceptual identity, no situation is experienced without it. The mind sees its conceptualized idea of the world more than it sees the world itself.

This experience pulls our species further and further away from what gives purpose to life. The very feeling of being alive which has become increasingly lost to humanity. This is the cause of any dysfunctional psychological patterns we encounter within ourselves, as well as others. The mind demands so much of our attention, that unlike any other species on Earth, we are barely able to feel the deep inner aliveness within our bodies, the source of inner peace as well as the source of feeling purpose in simply being alive. Which aids our ability to find purpose in many simple or complex situations. The dead end job that the mind declares to have is simply a job, no more or less. It is the need to find its specific brand of suffering in order to hold an illusory identity together that brings in negativity. If we become ready to see this from a deeper place within, the same job can become a beautiful practice of finding peaceful aliveness within the simple tasks. Any routine in life can shift from mediocre to this one beautiful moment that is here now. Instead of watching the clock to see how much more dreadfulness is remaining until we can clock out for the day, the value of being fully here now, the only place life ever exists, can become increasingly meaningful. If we decide to find a new job, this can become less filled with stress as well. The ability to be at peace with the tasks of the so-called dead end job will create more likelihood that we find a better job more suitable for us, instead of always feeling like the world is against us.

As a kid it was not yet possible to see how thoughts and emotions feed off each other in an unconscious way. To notice how a thought would trigger an emotional reaction that would trigger more reactive thought patterns. As far as I could tell, it was simply "me" thinking. It wasn't until early teens, when my thinking became lost in patterns of self loathing, that the first clues started to present themselves, something here is amiss. I was told I had depression, although at that time, this was as deep as I could understand these experiences. A symptom, a mental label that had very little use in a deeper understanding of what causes one to be "depressed". Not that I was falsely diagnosed, just that I was diagnosed based on the level of understanding others could provide me with. Other humans that only know themselves as deeply as their own thoughts understand, and through the mental education they achieved. Nothing wrong with the amount of understanding they had, just that it was limited for the subject we were working with; thoughts cannot see beyond thinking. The best they could do was treat the symptoms with pills and have my mind talk, or think more about this experience known as depression. However, this was not a failure as much as it created a need for deeper questioning to come in that needed time to look for deeper understanding.

Fortunately, when it comes to symptoms like depression, what we resist will persist. At least until the steadily increasing inner resistance forces us to look more deeply within and become more conscious of ourselves. Provided one becomes ready enough through this process, there becomes no other solution but to look more deeply within than the mind's identity, look deeper than the resistance (ego) itself. Which requires a certain amount of acceptance towards what we might find. From this deeper level, no longer fully identifying with the phenomenon, we can face inner resistance more directly, more willingly, and begin to find a level of inner acceptance that is deeper than thought, and needs no thinking about it to experience. This begins to transform our inner resistance into inner acceptance. We begin to find as our overthinking slows down, we gain access to a deep peace that was always within, just covered up by continuous thought. An inner peace that is more who we are than the conceptual thoughts about ourselves can comprehend.

This peace is already within you, maybe you just haven't realized it yet due to an overactive mind, yet there is nothing wrong with this. It takes an overactive mind in order to notice it enough for the ability to step back from the mind, metaphorically speaking, and come to the realization that you are more than the thoughts the mind is experiencing, you are the awareness of your thoughts. Allowing the physical awakening out of identification with thought to occur, and feeling the full aliveness within the body, the feeling of Being. Allowing a deeper level of consciousness to come into your life and into this world.

So far in what is the very beginnings of an evolutionary shift in humanity's consciousness, the most likely to begin awakening out of ego are those who have suffered enough to be able to see it more clearly. As it takes a certain threshold of these repetitive patterns before we begin to see them from a deeper place within ourselves and begin to move beyond them. Yet as our evolution moves on, less and less people will need suffering in order to help them become more aware of themselves. Society will become more knowledgeable, as well as the numbers of humans who have awakened will help people discover their deeper selves without as much mental and emotional dysfunction to be necessary. This being said, you are at a very important time within this process. You are among the earliest of humans setting the stage for the mass shifting to take place. Therefore, no suffering is without its purpose. Any suffering that helps any human find their inner peace is helping create this shift for our species as a whole. Any suffering that leads to peace has truly fulfilled its purpose.

Growing up, our family, friends, and our overall culture/environment teach the mind how to think and how to create patterns of suffering in a way that is familiar to those around us. We are all responsible for our actions, but it is no one's fault for how they act as well as what they influence on an unconscious level. If they knew what they were doing, they wouldn't be suffering, and wouldn't be spreading, or influencing suffering onto others. If you understand what these words are pointing towards, you are likely ready to move beyond your own unconscious suffering, and find the peace that these patterns have been covering up. They can now be your guide for finding deeper self understanding. The curse of suffering becomes transformed into a gift that points us towards a deeper peace within, a deeper knowing of ourselves.

The mind wants to think, it wants to have our constant attention. A simple impulse that has built up the momentum of overthinking for many millennia, most likely stretching back to the earliest hominin species. Slowly but steadily bringing us to the point where we are today. Consciousness within humans has become so confined by the mind, that we have nowhere to go but beyond this level of consciousness, beyond identification with thinking. Not without thought, but no longer bound by every thought. Bound by the mind's unconscious need for negativity, judgmental, and defensive ways of living. No longer dragged into many different directions in many unconscious, and often chaotic ways.

We might be the most evolved species on Earth. If that's true, it doesn't mean we are at the end point of our evolution. The state that humans have put our planet in today would suggest that we may not be as evolved as our thoughts claim to be. We may not be as conscious of our actions as we think we are. When something is lost, and then found again at a much later time, a newly found appreciation gives it a level of depth that was not there before. The feeling of aliveness within our inner Being that other species experience is not as direct or as conscious as it becomes when a species is lost from it, or at least without knowing of it for as long as humans have been. When we are able to find it again through this journey of first gaining complex mental intelligence, becoming lost within thought, and eventually moving beyond identification with thought. Awareness of consciousness itself becomes deepened, or purified in a way that other species that have not taken our journey can know.

What the mind would call suffering, is creating a depth within you, without the mind's ability to realize this is what is happening. This depth will allow you to experience yourself/life from a much deeper level than the surface reality of thoughts and emotions. The challenges we face today, both individually and collectively around the world, are what humans need at this time in order to see beyond our current level of consciousness. Because if life was without any challenges, there would be no reason to grow, and evolve.

Once we begin to see how our own mind can create problems for itself, we have become ready to move beyond the ego level of awareness. There is nothing that needs to be done, there is no place to get to. The depth within you is already here now, all that is needed is your awareness within the present moment. Once we become aware of our own egoic patterns, they can no longer confine us within an illusory mental identity. The mind will think about its own repetitive reactive patterns, as well as other peoples, and there is a place for this. However, it is more important to notice our own reactive patterns than it is other people's. It is more important to notice these patterns at the moment they present themselves as opposed to thinking about them before or afterwards. The present moment is the only place where transformation occurs. Because it is the only moment we ever experience in life. Thoughts are time based, reality is here now.

The belief that consciousness is a product of the mind as opposed to the mind being an expression of our consciousness seems to speak for our attachment to the mind more so than it holds a truly tested scientific understanding. This belief structure seems to hold science back from asking deeper questions about what consciousness really is. There are plenty of answered questions about the mind. Many advancements in health related research have been made for example. However, the scientific method can only find answers to the specific questions that are asked.

Until more humans that might label themselves as scientists begin to understand themselves on a deeper level than their thought identities, our collective scientific discoveries will likely be limited by the general belief that everything we perceive is through thought. That it takes a thought in order to perceive anything consciously. Overlooking the possibility that it takes a conscious being in order to have a mind that thinks, there are many species on this planet who don't have a human brain, and are able to be conscious without continuous overthinking.

Thoughts about what consciousness is might keep the mind awake at night trying to conceptually understand just what being conscious really means. Fascinating, complex, extremely brilliant conceptual portrayals of imaginative ideas that are mesmerizing, and fun to play around with. Yet none of it is the deeper truth, if the deeper truth is still unknown. No thought can grasp the true depth of what you are on the deepest level of your Being, the living energy that inhabits this form, this body. You are not the form, which is paradoxically a part of what/who you are. On the deepest level, you are consciousness itself, a living presence within the present moment. If this was not true, you would not be able to read these words now, or perceive anything through sense perceptions.

As you become more aware of this truth, life will naturally become more peaceful. As all the unconscious needs for suffering will dissolve through becoming more aware of them. Not because of some new belief system. Simply through moving beyond the need to be attached to stressful thought patterns for the sake of holding together a mental identity. When we are content within ourselves, there is no reason to not be at peace with the world. As this book will explain in greater detail, the more humans that realize this, the less we struggle to create a less dysfunctional world, and the more naturally a peaceful world can be found.

You play an important role in the evolution of our species. You have the potential of becoming one more awakened human on this planet. You are the change this world has been waiting for. You are the universe becoming more aware of itself through the human form. Consciousness is rising towards the next step in our evolution. Consciousness becoming aware of consciousness itself.

Another question that could replace the chicken or the egg in order to momentarily stop the mind's thinking could be;

What is aware within a silent gap between two thoughts?

Evolving Exercise

Take short moments throughout your daily routine to simply pause. Be more deeply where you are now. Simply notice your thoughts and emotions as they are, without judgment of how they should or shouldn't be. Bring attention into the body by noticing your breathing. Notice the feeling of aliveness within the abdomen, as well as other parts of the body. This is also very helpful at any time you feel overwhelmed by thoughts or emotions. Bring your awareness to the simplicity of the lungs filling with, and emptying themselves of air. This practice allows more space to come in between thoughts. More space to come into your life. The heart beats on its own, many organs work without our knowing. However, the lungs allow a certain voluntary participation that helps us become more conscious of the inner body.

As this practice continues, you will more easily find this living energy within the chest, as well as increasingly throughout the body. As the energy of Being is always within you. It is the deeper you that is beyond the thinking mind. A certain alertness is required to notice the subtle yet profound energy flowing throughout your inner body, or inner Being. The mind cannot grab our attention as easily when we focus it within the body.

Gaps between thoughts may be extremely short at first. However, the more you can practice taking a few moments throughout your daily routine to simply notice the subtle beauty of your breathing, the longer these gaps between thoughts become, and the more peaceful life becomes in general as the mind can no longer drag you into every direction that is suggested.

There is no way to directly stop the mind while its momentum is still strong. Yet the more moments you take to simply practice bringing attention into the body, the more the mind slows down on its own. Without your continuous attention, overthinking cannot exist. What might start out as a tedious exercise to try and achieve peace, will likely become more and more of a daily celebration towards all the subtle beauties in life which the overactive mind cannot help but overlook. You will become more aware of how thought is only a small aspect of who you are while becoming more aware of the vast realm of intelligence that exists beyond thought. Together this enables the ability to see first hand how all the truly meaningful things in life, like joy, love, beauty, and inner peace, all arise from beyond the realm of thinking.

THE PRESENT MOMENT

"Welcome to the present moment. Here. Now.

The only moment there ever is."

-Eckhart Tolle "The Power of Now"

There is no time in our lives that is outside of the present moment. Most of our thoughts have to do with our memories of the past, or some mental projection of the future. Conceptually speaking, time is important for many practical purposes. Thoughts about time can be one of the greatest gifts of having a human mind, yet most of our thoughts of the past and future are simply for the sake of continuous thinking. Holding together the story of "me" while keeping most of our attention away from where we are now, where life is always happening, the present moment.

If you look around you at this moment, there is no past or future here. There are no complex stories of the past, or complex projections about the future. Outside of thinking, this moment simply is as it looks. Our level of inner suffering is always dependent on our level of resistance towards this moment. There is never any true stress within this moment until the mind proclaims that it shouldn't be as it is. There might be a situation that we don't prefer, as well as situations that call for action, but it's the resistance towards what is within the present that turns the simplicity of what is here now into resistance based suffering. The mind essentially proclaims its resistance to what is within this moment to be more important than the situations we encounter. Not that our thoughts see it this clearly, the mind simply sees what it doesn't want.

No situation in life is truly a complex problem until the mind's thinking has imposed the conceptual label of "problem" onto the current situation, reinforcing and continuing familiar reactive patterns as the story of our conceptual self. The mind might think, "This always happens to me", without realizing it's more so the repetitive reactions that are always recurring. Every moment is simple without these complexities of the mind. Every present moment, and situation within, is just this one step here now. Some more challenging than others. Yet life becomes much simpler, and much more beautiful, when the mind no longer drags us away from this one moment into its repetitive resistance patterns towards what is here now.

As we carry out our daily routine, the mind is constantly in a need of escaping this one and only moment, as it is an amazing tool for doing so. Imagination can be a wonderful thing with many useful purposes, yet can equally be a burden onto itself. There are many times throughout the day that the mind might tell itself it needs to get to some future moment that is seen as more preferable. It is repetitively thinking about how much it wants to avoid certain future situations it does not want to experience, or doesn't want to have to deal with. As well as randomly clinging to its memories of past events in a way that repeatedly tries to bring an identification with these memories into the now while unknowingly creating stress through the clinging to what no longer is within this moment. "If I could only be the me I was then", or "have the life situations of that time", thoughts repeatedly tell themselves. Much of our thoughts can use most of their energy avoiding the only moment we can ever live within through a need to be there instead of here.

Our thinking can also often feel as if life should be filled with a series of montages in short lived clips that are carefully edited like in a movie which could allow us to get to the perceived more important events, all the "good" stuff without all the so-called arduous steps involved with getting from here to there. The issue with this is not just the need for a time machine, but the need to miss about 99% of life itself. An ingrained need to completely overlook the beauties that lie within the simplicity of allowing this one moment to be as it is. The world around us, as well as within us, becomes much more beautiful, much more joyful when the mind no longer continuously drags us away from this one moment into a repetitive narrative of how ugly everything is, or how dissatisfying this moment is according to these types of thoughts. The simpler we can allow this only moment to be, the less our mind can claim its story of complexities that it can't help being lost within, simply because the mind wants to think continuously. It demands attention as much as possible. In doing so, the only moment we ever experience needs to often be too much of this, or not enough of that. When we simply accept this moment as it is, we are no longer lost in the mind's drama of how it should or shouldn't be. Allowing a deeply peaceful feeling to arise from within us in between the stream of thinking. This is when true presence emerges into this life, the beginning of knowing ourselves more deeply than the conceptual thoughts about ourselves. Allowing our thoughts and actions to become less chaotic, and more balanced.

As the current normal way of living moves forward, the mind rooted in peace continuously reestablishes what could be called the heaviness of time. A conceptual burden that can feel like the weight of the world is on our shoulders, the weight of what we should or shouldn't be doing with the little amount of time we supposedly have on this world.

We do only have so many years in this life, but the extremes that the mind can create for itself is not so much for the sake of acknowledging this, as it is for the sake of holding together the identity with this familiar relationship towards the heaviness of conceptual time. Just as heavy can be the thoughts of how long a certain situation is taking, or how we have not yet achieved all the things we wanted to. In one way or another, the story of time becomes like a weight that the persona must carry around everywhere it goes, for the sake of identifying with thoughts about time. This is purely conceptual, but can literally be felt within the body as a gripping heaviness, something the emotions cannot help but cling to in a resistance based way.

A time based persona becomes something that must be maintained, as our perception of time holds the mind's illusory sense of self together. Keeping us in the one place we can never truly know ourselves, never truly find liberation from the heaviness of conceptual time. Practical planning is important, but within this heaviness, no view of time is without some degree of suffering. It could be as simple as thinking about something you need to do after work, or something you didn't do several years ago. Simple future, as well as past actions can easily become tainted by carrying this burden of the time based self around. Worrying about a future event not going as we want can pull our attention away from the more important task we are trying to accomplish now, more important simply because it is the step we are on now. When we give this moment our full attention, the heaviness crumbles, and every task can become peaceful, without heaviness.

Life is always one step at a time, but with such a grip of the past or future always lurking in the background, every step can seem like many steps all at once, never quite getting to where we need to be, rarely being able to experience where we are now. Always carrying the story of time as the story of who we are, or who/where we want or don't want to be. The only way out of this heaviness is by first noticing it when it appears within the mind and emotions, then simply realizing that there is no place you can get to that is outside of the present moment. When it was the past, it was the present moment, when the future comes, it comes as the present. In this sense, there is no time. Learning from the past and planning for the future is very helpful, but even that can only take place within this moment. The weight of conceptual time will only dissolve when we no longer lose ourselves within it. Be here now and there is no heaviness to carry, no more crosses of time to bear so to speak. What a relief it becomes when we only need to be within this one moment, the only moment we can ever be in.

Boredom is a great example of a form of restlessness that is created by the mind's need to be there instead of here. Many times it is a general need to be doing something else that is surely more important or more entertaining to the mind than what is here now. The mind has a need to always be busy doing something. "There is no time to be here now" it tells itself. In truth, no matter how busy we may or may not be, it is always beneficial to also be more fully here now. More fully noticing the feeling of the inner body, as well as noticing the environment surrounding us. Whenever the mind is certain it should be there instead of here, it creates and clings to some level of suffering within, some amount of anxiety. Always reaching towards what does not exist within this moment, creating some amount of tension which covers up the peace that is always here now.

Within continuously reestablishing the idea that life is only in the past or the future, the present is just a means to an end. This creates a relationship with life that forces us to see ourselves as separate from life. Some kind of anomaly of life that is alive but also somehow paradoxically not a part of life at all. Always at odds with the ways that life is being experienced, at odds within the concept of "me against the world". Simply because the mind cannot find any deep connection with what life truly is as a conceptual thing, as a thought thinking about thoughts. What we truly are is consciousness experiencing what is within this moment, it can be said that we are the present moment experiencing itself. We can experience exciting and captivating thoughts within this moment, but something deeper must come in for us to realize the depth that connects us with all of life. To be able to experience first hand the feeling of Being within ourselves that can then also be seen within other life forms. All forms of life are different forms of the same one expression of life itself, the present moment coming to life through form.

There have been many scientists who have made amazing discoveries because of their ingenious thought capabilities, but in many cases, the only reason for their "Aha! moment" could be described by a moment of space between thoughts, allowing something new, or deeper to come in. Most scientists with breakthroughs like this do not likely become awakened from identification with thought, but whether acknowledged or not, they benefited from a momentary gap between thinking that allows a deeper realm than thought to help their work. It is really only one step further to know ourselves more deeply than thought, and see the difference between our deeper self, and conceptual time, or time based sense of self. Once we no longer look for ourselves in conceptual time, the heaviness of time is lifted, and the mind is able to be used with less and less dysfunctional characteristics.

Today, curing boredom can be as simple as staring at our phones or other electronic screens that can do truly miraculous things in comparison to the technologies of the past. Yet are at many times used to entertain the overactive mind so it does not get too bored with what is here now, or merely out of habit. Becoming lost in what is an extension of our minds, rarely using these devices in ways that enhance our knowledge and understanding of the world, as well as ourselves. They tend to mostly be used as a distraction, or form of escape from the one moment the mind always seems to need to get away from, the now. That being said, there is no wrong way, or right way to use technology. One of the most beautiful elements about these devices is the large variety of things they can do for us.

On the other side of the spectrum, if you can notice that the mind is addicted to these devices, you now have the opportunity to move beyond the need to overuse or misuse them. These technologies have opened up many opportunities for us to live in beneficial ways that our ancestors could barely dream of. Yet, as long as humans are trapped within the ego state of consciousness, these technologies will mostly reflect this state. The less ego we inhabit, the more beneficial our technologies can become to the extent that we can barely dream of what kind of benefits we may encounter through our technologies in a world that has gone beyond humanity's Ego Stage.

The ego needs its proclaimed individuality, its time based self, as intensely as it does in order to survive as an illusory sense of self. The ego needs to overlook the deeper truths that would end the mind's unconscious obsession with time. Deep enough within us all, we know the truth that there is no time, only the present moment. The mind unconsciously needs its time based suffering in order to define its identity. It needs its heaviness of time to hold the dream together. It needs its suffering of continuing to avoid its suffering, which only creates and sustains more suffering. Time based thought patterns cannot last within the light of our deeper inner presence. As the joy of simply being here in the now arises through acceptance and feeling the inner body. The past dissolves within our deeper presence of the present moment. A thought can no longer be seen as reality, but only a possible pointer towards what is real, once we become present enough. A new level of consciousness is beginning to arise out of the thought/time based level of consciousness. Within an increasing number of individuals, stillness within is beginning to replace the restlessness that most humans cannot help but be lost in today.

Complaining is one of the main activities that the mind uses in order to avoid the present moment. Any situation that the mind does not want to face becomes an exercise for strengthening ego. Within every complaint is an imagined superiority through the exercise of playing the victim. The victim role is the best way for the ego to strengthen itself. There is no way to convince the mind otherwise when it has decided that this expression of suffering is absolutely necessary or absolutely reality.

Usually, it is just as important that others know just how wronged we have been, allowing the narrative to be repeated, and enhanced by other people's reactions. There is nothing wrong with this, all unconscious behaviors are a part of our process of becoming more aware of ourselves. It is through simply noticing these patterns within ourselves that is how we bring presence into the mind's reestablishing of the victim role, dissolving the illusion with truth. No complaint is truly necessary without a need to strengthen the ego. A need to resist what already is or what might come. If we are not content with what is within this one moment, we can always try to change what is, but complaining is like ingesting poison for the sake of being poisoned, not for the sake of changing what is.

There's no way that life can always go exactly as planned. Nevertheless, to a mind that has not yet been made conscious enough to see the futile nature of resisting what already is within this moment, creating suffering through complaining is the most logical conclusion. Like any other unconscious activity, the only element that is needed to change these patterns is to simply notice when they occur within the present moment. Thinking about this can help prepare you for when patterns need to be recognized. However, noticing and accepting these patterns within the present moment is the only way to transform inner resistance into the awakened state of deeper inner peace, or presence awareness.

This doesn't mean we must say yes to everything, but simply notice the amount of resistance that might be within our decision making. Simply notice the amount of resistance towards what already is within the now. Accept what is within this moment fully and suffering has ended within this moment, allowing clarity and patience to replace impulsive and dysfunctional reactions. The same reactive patterns will likely come back again and again until they can be fully dissolved or made fully conscious within the present. They can have a lot of momentum from being practiced for as long as they have been, yet acceptance has the same effect every time. No amount of inner resistance can survive your full acceptance of it. Even a small amount of acceptance will alter the experience. If you cannot fully accept what is within this moment, accept that. Once these patterns are seen for what they are, ego, it is the beginning of the end for the ego, and its dysfunctional suffering. Awakening is not a straight line, or in other words, there are present moments where we are a little more or less present, but on a broader scale, there is only one direction. We can only become more present as the present moment moves on.

A great example of the mind's natural resistance to the present moment could be a car trip with family. When I was growing up, I lived in Columbus Ohio with my mom and older sister. We would often make trips to see my grandma who lived near Dayton, about an hour and a half drive each way. I remember enjoying the view for short periods of time, lots of corn fields. What I also remember is the restlessness the mind would create within the "are we there yet !?" impulse feeding off a situation that could not be under my control. We couldn't be there if we're simply not there yet. None of us were overly dramatic with each other thankfully. Yet, that tension of needing to be there instead of here, was always prevalent to some degree while simply sitting comfortably in a car seat. Nothing was needed except the mind's demand of being somewhere that was not here.

Of course I was also a kid, and learning patience is a part of growing up. That being said, how many adults today are unable to simply sit and be content while in a waiting room or driving to work? Simply because their mind's repetitive thinking patterns have formed and continuously reestablished the opinion of "I shouldn't have to". Thoughts like, "I shouldn't have to wait this long"! "Why did the doctor set this appointment just to make me wait like this"! "Every day, traffic, traffic, traffic! I shouldn't have to put up with all these crazy drivers!" Drivers who are probably thinking the same thing, as their need to be "there" makes their driving more chaotic than if they could simply be at peace with where they are now.

An incredibly powerful source of freedom comes from the simple ability to wait without the mind's restlessness needing to be there instead of here. Allowing a world of inner peace beyond ego to open up within this one and only present moment. Any time we are in a situation where waiting is all we can do, the opportunity to practice bringing more presence, more peace into our lives arises. Bringing in more presence can take time practicing being present, yet this becomes increasingly worth it as situations that require waiting no longer drag us into a heavy, dense negative mood where no inner peace can be found, simply because the present moment cannot be found. Only the impulse of needing to get to the next moment and the train of negative thoughts that absorb our attention and our conscious energy. What could be a moment with a certain joyful simplicity within it, is instead transformed into a nightmare of overthinking that cannot help but resist for the sake of resisting.

An ego needs to find some kind, usually many kinds of conflict in life in order to sustain its illusory sense of self. Sustaining the mental/emotional patterns that are familiarized, sustaining the heaviness of time. Plenty of conflict can be found in the mind's need for comparisons. Always comparing what it defines itself as and what it declares itself as not. Comparing what the mind likes with what the mind dislikes, its favorite this, its least favorite that. Noticing many differences between ourselves and other people. Creating a source for a kind of superiority/inferiority to take us over. "I would certainly never dress the way that person does", could be one example. This doesn't mean we must renounce comparisons in order to go beyond our ego. Like any egoic pattern, simply by noticing the mind's attachment towards comparisons dissolves the ego elements involved. Ego patterns cannot survive your presence of them. Comparisons are neither good or bad, yet any thought that holds a certain heaviness is all we need to notice. If there is no heaviness, no restlessness, it is not ego. If a heaviness is noticed, this is simply because you are moving beyond ego. Madness cannot know that it is madness, to notice suffering within for what it is, is moving beyond it.

Many egoic comparisons will inhabit some degree of negative energy that the mind is creating and sustaining through its constant need for comparing, and/or criticizing. If we begin to see our own ego as the foe or enemy, this is just another mental comparison of ego. Creating yet another conceptual layer for the mind to defend itself against. Defending the conceptual "self" is what ego needs most in order to survive. Holding an illusory self together through conceptual comparisons. There is nothing wrong with this, nothing needs to be done in order to awaken from this state. Simply noticing the patterns dissolves the need to identify with them. Even if they must repeat themselves for some time, the general awareness of ego patterns means that they are dissolving within your presence now. Every story of ego is time based, and time cannot survive within the present moment. The more we are here, the less we are there.

An ego gains most of its strength from comparing itself to the "other". A mental concept of what the ego simply sees as "not me". Strengthening its identity through thoughts of how superior the mental "I" is towards others. Also, just as importantly for the ego are the thoughts of how inferior the mental "I" is towards others. Every unique expression of ego will have various degrees of both high and low egoic self esteem. As paradoxically, the two are one in the same. The mind cannot go for too long on one end of the spectrum without diving into the other.

Of course, on a deep enough level, there is no true other. Only on the level of forms. The only level of life that the mind can analyze and conceptually compare its illusory sense of self or story of time with. Separating itself from/overlooking what is essentially the same one consciousness that is being expressed through trillions of different forms. The truth is that this one world is one superorganism in many ways. These surface differences are all important, on the surface levels of life. Very important for all the practical aspects of living as a human, or any other species on Earth. Except that while the mind thinks about the importance of physical elements in life. It can not help but overlook the importance of knowing ourselves more deeply than the forms of life. The importance of knowing ourselves more deeply than the roadmap of conceptual time. Locking ourselves into the prison that is the heaviness of conceptual time.

Outside of the mind's thinking, there is no time, and the mind's use of time easily becomes dysfunctional when we mistake its story of time for who we are, and are not able to acknowledge the timeless depth that is within us all. The form you inhabit will age, and at some point die. Yet the deeper you is infinite. You are the present moment living through a human form. When the form expires, the deeper you still is as it is. Much like this very moment simply is as it is regardless of the countless numbers of moments we experience throughout our lives, it is still this one moment. Your consciousness is one in the same with the present moment. Regardless of the forms that temporarily make up this moment, the deeper you is infinite, this is the present moment on its deepest level. It is ancient, and brand new at the same time. Your presence is ancient, and brand new at the same time. Despite all of the changes that occur throughout time, your presence is always here now. Your very presence is what brings light into this one moment.

The surface aspects of life, like the different forms we inhabit as living Beings, are only a tiny aspect of our true, or deeper reality. When we are unable to know this truth first hand from deep within ourselves, the surface differences seem more important than they actually are. Causing dysfunctional, or even chaotic thoughts/emotions/actions with a need to defend a surface reality as if this is all that we are. As it is all that the ego can understand itself to be. Because the ego sees itself as a separate entity from life. It must defend this delusional understanding often, and at all costs. Making it easy to blame "others" for simple reasons, and especially for what we are not yet ready to see is occurring within ourselves.

Without the need for an extreme conceptual labeled "other", there could be no inner conflict towards other people that leads to all the different inward and outward expressions of madness that we see around our world today. There is a surface need to know the differences within the world of forms. Yet on a deeper level, if a human is truly content with their existence, or presence within this one moment, the only moment there ever is, there is no need to compare one's physical place in life, or conscious state with anyone else's. We can still find ourselves envying someone else's path/position in life. As well as being concerned for the choices someone is making, concerned about where someone else is in their life.

Yet, the unconscious need to judge, blame, or criticize is what fades away. A need to compare where we have been, what we know and understand, where we are, or where we might be headed compared to someone else's journey. Is no longer needed in order to feel good/bad about ourselves. Also no longer feels needed in order to engage with other people. Becoming more conscious of ourselves means we no longer need to use comparisons in order to sustain our judgments of someone else through our unconscious judgments about ourselves. Allowing true compassion to come into this world. Unconditional compassion for ourselves, that is, therefore, unconditional compassion for all living beings.

The need for a conceptual "other" was at one point an extremely successful survival strategy. Not without its side effects of creating madness for ourselves and others. Just not yet at a point where we can acknowledge these side effects as clearly as we can today. In many cases because of the technological breakthroughs of the modern world. Allowing us to become more aware of how similar we actually are with each other. In the past, most human awakenings out of ego have been only through seemingly random individuals, many who are likely unknown. As it is very difficult to describe the silent beauty of inner peace to an overactive mind.

41

Words become extremely limited when trying to point beyond words.

Some societies in the past have had various degrees of cultural awakenings of more than just a small number of individuals. Only to come and go within a world that has so far always favored the egoic expression of survival of the fittest. Meaning, given enough time, in some way or another, usually through violence of some degree. The population of humans lost within ego have always prevailed over smaller groups, as well as individuals, that have seen beyond the madness of the mind. Cultures that were either consumed by the madness of an externally invading army, or consumed internally as the individuals became reabsorbed into ego by their own overthinking minds. The more the Cultural Ego becomes involved in any society, the less present the collective individuals can be.

There have been people from the past, some that would later become known as religious figures. Who have had some success in helping point the way for others. In many cases only to have their message swallowed up at a later time by an egoic misunderstanding of what they were pointing toward. Turning the message of how to find the infinite peace that has always been within us all, into a way of worshiping a conceptualized version of a perceived external and egoic entity. Also turning the belief structures into some form of egoic hierarchy. A reflection of the ego stage of consciousness in many ways, one being its feelings of being separate from life itself. Conceptually worshiping some form of a God, or Gods that the mind believes will protect us if we worship correctly. Yet also seem to judge us in ways that make the most sense for an egoic mind. Turning what could be a pointer towards inner peace within this moment. Into some projected fear of time, as the mind can't help but impose an extreme importance on the ego's fear of death.

Resistance towards death can easily over-emphasizes the mind's concept of good and bad. Creating the worrisome need to be a good enough person within the eye's of the Lord in order to reach the correct infinity that can only occur after life. In some cases, dissolving any need to teach the importance towards being here now. As the biggest goal in this life is always in the future. Turning many religious teachings, what could be a lesson about the values of serenity towards what already is within this moment, into a need for some level of self judgment, a need for inflation of ego. The less conscious or less present humans become, the more resistant we become, and the more likely we are to literally create Hell on Earth for ourselves and those around us. As we are never fully here now, always aiming for some imagined future utopia and resisting the imagined Hell within the unwanted afterlife. Always avoiding the only place that heaven can be found, within ourselves at this moment. By overly avoiding the mental concept of death, we overlook the true purpose of life, being alive right now.

The major religions of the modern world could be viewed as being only based on mental belief structure, or only ego based. Which can make them look from outside observation as if their only function is to be dysfunctional in many ways. While it has been true that humans who are lost in ego have used the belief structures of Judaism, Christianity, Islam, and many other religions in an unconscious fashion which strengthens expressions of egoic madness in various ways. This has more to do with the ego unconsciously looking for anything that might strengthen its conditioned identity than it has to do with these religions themselves. The ego cannot help but turn religion, much like any other subject, into a reflection of itself, a reflection of its own mental identity. Creating God within its own image, creating the same byproducts of madness through the religion that it creates through so many other things.

It is also important to note that not all egoic understandings of religion create unconscious side effects. Religion can be a source for balance, humbleness, and deeper understanding within our lives as well. A human being can only go as deep within themselves as they are consciously ready, and the fully mental version of spiritual understanding is simply how the mind views spirituality. Thought based and not experiential in the sense that knowing ourselves more deeply has the inner body's feeling of deep inner peace along with it. It is not right or wrong to hold a religious belief structure. It is simply the understanding of higher power or collective consciousness from the ego level of consciousness. Not without some amount of dysfunctional qualities of ego identity, but also not without some amount of deeper understanding of ourselves. Many people within the ego stage can find great wisdom and guidance through religion. You could say that the amount of inner peace anyone at any level of consciousness finds through religion depends on whether we are looking for God through the mind, or looking for God through the heart.

Religious expressions with various amounts of egoic madness are only a symptom, and not the source for the inner madness of the human. Fortunately, most religions are rooted in wisdom, and can be a source for serenity as much as they can be a source for strengthening ego. When the ancient and timeless stories of good versus evil can be transcended into the understanding of finding our timeless light within the mental darkness. Any mythological story can hold various amounts of deep wisdom within them. When we see them as absolute truth, they tend to keep us from finding certain truths within ourselves. When the stories are not believed to be 100% true, when we don't identify with a story as a part of who we are, we can more easily find messages of wisdom in various places within the story. Historical Jesus, for example, was attempting to point this out to the people around him. Yet, if the general public is not yet ready to look deeply enough within themselves, a new way of seeing that looks beyond ego will likely become a reflection of ego within a short time. These patterns repeat themselves until they are made conscious enough.

However, throughout the history of Christianity, like most religions. The collective understandings have swayed back and forth between pointing more deeply within, to pointing more towards the egoic qualities of the belief structure. Swaying back and forth from pointing towards the importance of understanding the infinite, formless presence within us, that is more what we are than the form we live through. Towards unconsciously over emphasizing the importance of form. Through the ego's overly fearful resistance based relationship with death. Fear of losing the physical form it identifies with.

Death can be viewed as one of the most sacred points in life. It has the potential to teach us how to live in serenity, to be humbled in many ways through the situations we encounter in life. If it is not approached with resistance, or avoided entirely. There is no life without death. Many people today place the elderly in facilities as they get closer to death, pushing them away from us. Some of this is helpful and necessary for where they are in life, as well as their needs for staying comfortable. Unfortunately, sometimes it is simply to take the burden off of their family member's minds. Like sweeping it under the rug so to speak. Death in today's society can be easily viewed as the worst thing that could possibly happen, so it becomes avoided as much as possible. When someone does die, this triggers the impulses of the ego's resistance towards death. Everytime someone passes, the ego goes through some degree of avoiding its own death all over again, on top of mourning someone else's passing. Tainting the experience with negativity, and adding to our egoic suffering as we get older. Helping to increase the heaviness of time within the mind.

On the other side of this, anytime we mourn someone's passing with full acceptance towards what is within the present moment. The opposite experience occurs, it becomes a very healing process. Something that can connect us more deeply with our spiritual essence. This is why cultures that are rooted in wisdom celebrate death, and treat it as sacred. What we resist will bring more resistance, more ego. What we accept becomes joyous in a way. Not without the feelings of loss, some of which can feel painful. Yet accepting what already is can be very helpful for us to experience what is beyond the surface reaction based feelings. The pain on the surface is no longer all that there is. Giving the potential to notice the deep connection we have with the source of all life, the source of our Being. The universe is more deeply understood as the play of forms that it is. Forms come and go, but our spirit never truly dies, it simply returns to the source. We can't know this conceptually, but when we know our deeper presence, our deeper self. We find a deeper knowing that death isn't real beyond the ending of the temporary form. Our presence which inhabits this current form is infinite.

When someone is coming close to dying, they become more likely to transcend the ego before they die. If they find serenity towards what is, they find the surrendered/awakened state before they pass. Most religions have found this process sacred at different places/times, depending on the culture's current level of consciousness. The lesson of dying before we die, is to simply accept the present moment fully as it is, to accept the current life situations as they are. Within this, the ego is dissolved, its conceptual view towards death is dissolved. Conceptual time is dissolved; now every day is the only day of life, or at least the most relevant, regardless of how close we might be towards physical death.

With this, true joy is able to blossom into this world. The joy of the eternal now, the present moment, consciously experiencing itself as a human for a little while. No one needs to be close to death in order to die before we die, all that is needed is this moment now. Accept this moment fully, and there is no living ego, the illusion of time has been dissolved. If we're not fully ready, there is nothing wrong with this. The illusion may continue to come back in. You may need more time, time being present. Especially if the mind is still very active, but the ego is now on its deathbed. Acceptance towards the present moment is all that is needed. Practice being fully here now, and the heaviness of time is dissolving.

Most religious teachings contain some amount of spiritual wisdom. Many depend on the amount of wisdom that those practicing these lessons find within themselves. If someone is totally lost in ego, ego is all they will likely see. There is nothing wrong with this, we can only see the world from our current level of consciousness, and conceptual based understanding of god/collective consciousness is what we will see until we are ready to move beyond this level of understanding. The more aware someone is beyond their own ego, the more likely that some truth beyond conceptual understanding will shine through the words. Pointing beyond any identification with a certain image or form that the belief structures hold. Pointing beyond identification with ideology, breaking through any heaviness of mistakenly worshiping ego.

Luckily, most of our modern religions which have been translated and rewritten, retranslated, and rewritten many times. Still display pointers towards the original messages that were rooted in self awareness, rooted in pointing towards Heaven on Earth. Most religions don't have to be abandoned as we see more clearly how unconsciously they can be expressed. Instead they can be reborn into a deeper understanding that was always there. Simply waiting for us to become ready for a deeper understanding of ourselves. Allowing them to now become utilized as a tool that helps point us more deeply within the infinite depth that lies within ourselves. It is also not mandatory for anyone to stay within any one religion for the purposes of awakening out of ego, or for any other purpose, everyone's journey is different.

It can be much easier to see time-based mental/emotional reactive patterns in other people at first. However, until we become aware enough of our own patterns. Another person's ego will likely reestablish, or even amplify our own unconscious reactions. Inner resistance loves to share itself if it can. While this can also be very helpful in teaching us about our own patterns. We may find it helpful to limit our exposure with certain people lost within a deep heaviness of ego, when we can. At some point within the shifting process, this will no longer be necessary. A balance is also necessary to know when it is practical or impractical to distance ourselves from those who trigger reactive patterns within us. There is nothing wrong with where others are along their journey, it is not good or bad, it simply is as it is. For most people, resistance is needed in order to have the opportunity to move closer towards the next level of consciousness. Beyond the level of resistance based life, beyond the concepts of good and bad, beyond being lost in identification with mind. When the dream becomes increasingly like a nightmare, we become more ready to awaken from the dream of ego. So there is literally no one who is in the wrong place within this shifting for our species.

In most people, the process of becoming more present is filled with moments of not being present. As the Zen phrase goes, the obstacle is the path. Any moment we realize we are not present, we become more present through simply noticing. Transforming what is essentially unconscious into a higher level of consciousness. Usually a little bit at a time, but conceptual comparisons or scales are not very useful here, only this moment now. The process may take time, time being present. However, there can be no future plan or goal. Only the goal of being fully here now can help us along this path of becoming more present and dissolving the heaviness of time, and the overactive mind.

The mind has many repetitive thoughts which reinforce the time-based self, and continue unnoticeably until we reach a point of readiness to become timeless. Thoughts about "my life", where we have been and where we might go. What the ego calls "my life" is always the time-based illusory self. Not the historic self; all our life experiences have been a truthful reality that happened within the present moment. It is only the clinging to these past moments as if they are who we are now within this moment that makes them dysfunctional through reestablishing the past in the now.

The illusory sense of self is completely time-based, and would not exist without the way it continuously reacts with the past and future. Always needing to bring time into this moment. Because deep down the ego knows it would not exist without endless thoughts about time. In truth, there is no moment of our lives that is outside of the now. All past and future was and will be only within the present. The more we bring this understanding into our daily lives, the less our minds can drag us away from this moment into a filtered reality that must be defended at all costs. Becoming defensive over the smallest of issues if our thoughts see this as necessary. Notice how repetitive thoughts about the past recreate certain feelings of familiarity. The more we notice, the more awakened we become.

Death is the ego's biggest fear. In truth, there is no death, only the death of forms. When we become still enough within, we begin to find this truth first-hand. Not through belief structures within the mind. Through a deeper knowing that is expressed as a deeply peaceful feeling within the inner body. A deeper knowing that life and the present moment are one in the same. Life is infinitely within the now, it is your conscious awareness that brings light into this moment. As Jesus said, "You are the light of the world".

Illusions of time are the only thing that dies with an ego. Just as the form is the only thing that dies at the end of an individual's life. Ego is its own form in a way, made up of repetitive thoughts and emotions. Made up of a kind of worshiping of forms, and an obsession with forms of thought. Overlooking the formless essence that is much more deeply who you are. No matter what type of self esteem the mind is convinced it has, the victim role is an important one for the ego. Patterns that need to demand, "I'm being wronged", for this reason or that. Cannot help but create suffering regardless of the situation. Always recreating the need to leave the present moment. The need to be less conscious within this moment, reestablishing the same dysfunctional patterns to continuously prove, this is who I am. Continuously clinging to negativity to some degree.

On the flip side of this, as we become more present, the ego cannot help but give its patterns away. Since the mind is always demanding our attention. This need makes the ego more obvious as we become increasingly aware of the mind's thinking from a deeper place within. Allowing an opening for inner peace to become more easily recognized in any given moment. We begin to notice more clearly a pure lack of suffering from deeper within, a sort of Isness of this moment. A sort of spaciousness within the body, the opposite of the heaviness of stress. The present moment now holds a certain joyful quality regardless of the current situations. This allows us to become more accepting towards what is within this one moment. When we are at peace within ourselves, there is no need to blame the external world for our mentally perceived shortcomings. As we become more familiar with this deeper dimension within us. A deeper wisdom than our thoughts can begin guiding us through life. When actions are needed, they become more easily performed, from a more conscious level than the mind's overthinking.

When our attention becomes anchored within the now. It becomes impossible for the mind's resistance based thinking to pull us away. Unconscious negativity becomes less attractive as the deeper peace that has always been within, becomes increasingly inviting. Like oil and water, the two repel each other. When we know we are the presence within, we are no longer as easily pulled back into what used to be so unconsciously inviting. When others unconsciously resist, that becomes okay as well. This moment simply is as it is, and everything is as it should be. Therefore the most beneficial decisions can be made more easily by no longer clinging to an illusory sense of time, and the illusory sense that life is usually out to get us through the ideology that we are separate from life. The illusion is that intelligence only comes from thinking, the truth is that overthinking reduces our intelligence.

There are many social causes in which people may become a part of that can be very helpful in progressing and improving our society in its current dysfunctional expressions of ego. While many of these can be very helpful to society, no social movement can be more conscious than the people involved within. Meaning they can create positive change, but if they are still lost in ego, they will likely have to create some level of suffering for themselves in some fashion while attempting to create change. There will need to be a mentally labeled version of "victim" that they identify with. Regardless of whether or not there is physical oppression in which they are directly experiencing. When we see protesters on the streets today, shouting and holding up signs, they cannot help but also hold a strong negative energy which will seem extremely important from an ego's point of view. However, the less ego there is within the members of any particular social movement, the more it becomes free to be a form of acceptance towards how things can become, as opposed to over-emphasizing a resistance against the way things have been. Not without stating the dysfunctions of how things have been, but with less egoic properties, a social movement becomes more about moving the present circumstances into a brighter present moment. Instead of fighting the past for an imagined better future, which will likely bring more ego elements along with it. The more we can surrender to what is, the more likely we can create positive change, only in the present moment.

There is nothing right or wrong about our level of consciousness, it simply is as it is. Yet as we become more collectively awakened. Social causes for improving society will become less of a battle against this or that, and more of a movement of peace towards the change that is simply ready to occur at this time. A movement that is not consumed by egoic negativity will simply have far more energy and momentum to be used towards the positive possibilities. Any social movement that is aimed towards more peace and unity within this world is headed in the right direction regardless of the level of consciousness for those involved.

Being at peace with the way the world is today in order to create a more peaceful tomorrow can be tricky when we see all the social injustices that are created and maintained out of ignorance and how the majority of society is unable to change these behaviors. Mostly because change is like a form of death to the ego. Being a physical victim, and a conceptualized victim are two very different things. One is simply the situation within this moment just as it is, the other is a mental/emotional identity that holds itself together. Holding the pain together by reliving past situations as often as we can through negative thoughts and emotions for the sake of clinging to an identity of who we are.

The more we work to change a physical situation without a need to bring in the self-victim label, either for ourselves or others. The more powerful the movement can become. Mahatma Gandhi was only as successful as he was in helping free India from England's colonization because of how rooted the movement was in peace, in non violent truth. As opposed to being rooted in aggressive victimized exaggerations of us versus them. He was able to see deeply enough within himself to know that no matter how much he may have disliked humans labeled as English, they too were simply humans living on this planet.

Without a need to be violent to any degree with these fellow humans, India's movement was not as handicapped as England's violent reactions became. The worldview started siding more and more with the peaceful protests. Thanks to this simplicity of being rooted in the now, what was labeled as a huge problem by many, was able to be transformed back into the simple situation it truly was. It was simply time for England's colonization to end. In the end, both sides benefited in different ways. There is no life without challenges. How much would the average person's life improve without the need to over personalize, and label one's self as the victim of simple everyday situations? Not to mention no longer using the victim label when it comes to collective social situations. Amplifying the victim role within a large group of people fighting for recognition of the same label of victimhood as opposed to being conscious enough to create more peace within this world while speaking and marching for whatever the social cause may be.

There are many societal situations of injustice within our world today that can look impossible to change. In some cases, like racial discrimination for example, certain aspects may or may not be able to change significantly until enough of the collective human consciousness rises above the level of ignorance that has created these issues. This can easily make people feel hopeless in terms of the necessary changes that they may or may not see within their lifetime. Yet the one thing anyone can do, if they are ready, is no longer cling to the mental narrative that makes the situation that much worse. Living with true physical oppression is not an ideal lifestyle for anyone. Although, how different would the experience be without the repetitive thoughts and emotions that continuously remind themselves of how they should feel horrible all the time within the situation?

There would still be a strong need for physical changes to take place, but no longer fueled by inner suffering that the mind believes it must cling to as an identity. Not clinging to an identity of "the oppressed" allows simple solutions to be found more easily. As most of our energy is no longer lost in defending a mental image of the victim within the situation. Less aggression would be displayed within us, as well as towards other people. Life would still be limited in certain ways within any community being oppressed, and everyone experiences this in different ways, to various degrees. There would still be physical monstrosities in different ways throughout society that need changing. Yet a certain heaviness that resistance-based thinking carries around with it would no longer exist. Within this, a kind of grace emerges into this world out of past patterns of resistance. These patterns have created for themselves a new found serenity, as there was nowhere else to go in order to get away from inner resistance but to transform the resistance itself into acceptance. What was a curse has now become a gift of opportunity. Who knows what new world this new way of living might bring. Only when enough individuals become present, will we see a world that is rooted in wisdom as opposed to ignorance. This begins with individuals like you within this very moment. As you become more present, you bring a new found peace into this world, a new way of Being. Something this world has been patiently awaiting for a long time. The dissolving of identification with conceptual time, into the realization that there is only this moment.

> *"Remember that sometimes not getting what*
> *you want is a wonderful stroke of luck"*
>
> -Dalai Lama XIV

When I was 27, I found myself back in Columbus Ohio, sleeping on my mom's couch again, after a second failed attempt to move to California. I held the belief for several years that if I could simply live in California, my life would be great. That the problems I'd defined myself as having in Ohio, would somehow melt away in the California sun, or something like that. Unfortunately, I had very little understanding of the price of living and left Ohio with nowhere near the amount of money I would actually need.

I had also recently been diagnosed with epilepsy. I was experiencing random seizures while asleep at night. Likely produced by both genetic factors as well as a continuous build-up of stress patterns from adolescence through my twenties. I was rolling and falling out of bed many times the previous year or so, and had dislocated my shoulder several times from convulsing while hitting the floor. Yet I was convinced that this could only happen while sleeping. One of my cousins I was staying within California, Stacha, witnessed me having several grand mal seizures. One or two of them during the daytime, and luckily she had the courage to tell me I needed to go home.

What I thought was a huge failure, one out of so many others I would tell myself. Suddenly turned into me being at the exact right place, time, and state of mind to be handed the message I didn't fully realize I was looking for. A book that my sister gave to me, was suddenly pointing out inner truths that I may not have understood as deeply if I hadn't been failing as much as I currently thought I was on the surface level of life. Echart Tolle's "The Power of Now" was the first book I read within the spiritual subject that was not only able to very clearly point towards how the mind creates its own suffering. It was also able to point towards the deep peace that is always within this one moment now, in a simple and profound way if the reader is ready to see what is being pointed towards.

Eckhart's main theme of this book, as well as all his teachings, is that the present moment is the only moment that life consists of, as well as how overthinking makes us unconscious. This was the first strong glimpse I was able to have towards just how overactive my mind actually was like any spiritual teaching that holds true depth. The lessons didn't instantly cure me from the symptoms of ego. Yet the inner glimpse was deep enough that the shift had begun. From the position of trying to always get away from all the time-based inner suffering I was experiencing in life. To suddenly find a deeper experience of acceptance towards what is within this moment now. Not through forcing any sort of acceptance but becoming naturally ready to find peace within the surrendered state. The deeper consciousness within us that simply is as it is and has always been here now is only covered up by an overactive mind.

As I write this, I have not yet come to a place of full awakening out of ego. As the present moment moves on, I become more and more consciously rooted within inner Being, inner peace. Yet, the mind can still pull me into negative reactions to some extent from time to time. More importantly, since that initial glimpse. I have witnessed what could be described as a growth of inner spaciousness awareness. While paradoxically, the dissolving of dense inner resistance patterns within the mind and emotional body. This has slowly but steadily transformed habitual inner conditioning patterns of suffering into a joyful or blissful inner peace awareness through the practice of a simple acceptance of what is within this moment.

By gradually becoming more present, I have had the extreme privilege of watching a movement of evolution consciously shifting within this form. Like watching a sunrise, only you are the sun that is rising. You are the consciousness that is waking up from the night's dream. You are the universe becoming more aware of itself. A shift in which humans are individually engaged, but this is also a collective expression of humanity that is far deeper than any individual. So far, on the collective scale, this process of awakening is usually slow and steady for most people. Yet every awakening has unique properties, as every ego has unique conditioning patterns to awaken out of. Very few people will experience an instant shift, as Eckhart Tolle speaks of. Yet the more present we become, the more rewarding being present becomes.

Anyone can become fully awakened at any moment once the inner awareness becomes strong enough. In "The Power of Now", Eckhart tells his story of how he lived with extreme depression and negative thinking for a long time in his younger years. Almost committed suicide in his late twenties. However, suddenly, the thought, "I can't live with myself anymore," caused him to question, "Who is the "I" that can't live with "myself" anymore?" Suddenly, seeing very clearly that there were two different identities in the sentence. The mental "I" and a deeper "I" that was observing the thoughts. Instead of physically dying, he says his ego died instantly. Followed by a newly found inner bliss that was no longer being covered up by negative thought patterns. However, he says this transformation took him some time to understand what had actually occurred and even longer before he could put it into words for other people.

This sort of instant awakening is very rare and can require an extremely strong level of inner suffering, which in itself was a part of Eckhart's transformation, making it not as instant as it may appear. The curse of the mind's unconscious need to suffer is actually the first step towards moving beyond identification with thought. Like the old saying in Zen, "No mud, no lotus." On a deeper level, there is no time involved within this shift into timelessness. Only the recognition of the overactive mind and how much it holds us within the heaviness of conceptual time, away from the now. Often simply for the sake of the mind's obsession with endless thinking.

Self-talk is what Eckhart noticed when he suddenly became aware of thoughts in his mind from a deeper place within himself. Self-talk is another way that the ego is strengthened. How could the voice you hear in your head not be you? It is you, and yet it's not. It is electrical impulses in the mind, mimicking the sounds of your body's vocal chords into formulations of thought. Like any kind of thought, this can be helpful in many ways and can advance our lives in many ways. Yet, it can also create continuous restlessness in many ways. A continuous heaviness of time being repeated throughout the day, throughout the present moment. Continuously creating an addictive need to think about the mind's story of time. If you see someone on the street talking out loud to themselves, you may think they're crazy. Yet, the talking-out-loud aspect is just an intensified version of what most humans do throughout the day. Self-talk can involve practical thoughts the mind has throughout the day. Like the task you are performing at this moment, the next thing you're going to do, or where you're going tomorrow.

Self-talk only becomes an obsessive expression of ego that creates dysfunctional overthinking when the story of "me" becomes entangled with what the mind is thinking when we identify ourselves as the mind's story of self. Often turning short-lived thoughts into long stories about the time-based mental "I." Where "I've" been, or where "I" need to be in the future. Always imprinting the egoic version of ourselves onto every thought the unconscious mind is able to. Rarely allowing the present moment to come into our lives. More important is the need to think about our conceptual idea of life.

Like any expression of ego, the moment you notice it, it begins to give itself away. No time-based identity within thinking can survive our continuously growing awareness of it. When we notice the mind talking to itself about itself, this is not something we can simply tell ourselves to stop. The repetitive patterns are too strong to simply stop thinking on demand. Resisting will only create more fuel for more resistance-based thoughts. Yet, the more we can simply notice this with a certain amount of acceptance of ourselves and acceptance of the fact that these patterns will continue to present themselves for at least a short time. The easier it becomes to give these addictive patterns less and less of our attention. Less and less of our identity as practicing being present continues.

It will take practice to be able to anchor yourself within presence awareness. Yet all that is needed is your awareness and acceptance now. Awareness and acceptance are essentially one and the same. Simply being aware of unconscious resistance is conscious acceptance. There is nothing to do and nowhere to get to; only a certain alertness is required. It is the resistance that dissolves itself away as you become more present of these patterns.

It can be easy for the mind to ask, "Just how long is it going to take for me to become fully awakened?". This is another way for the time-based mind to create more time for itself, as it is an unanswerable question now. There is no future goal that can be set when what is needed is always here now. The true miracle is that it is happening now. After millennia of human suffering created by overthinking, you have broken through from that dimension of consciousness and are able to understand what these words point towards now. At this moment, you are awakening from the illusion of ego, the illusion of time. This process may take some time being present. Luckily, the present moment is all there ever is.

Perhaps more efficient questions than the amount of time this might take could be, "Am I here now?", "Can I observe the mind's thinking now?" and "Can I feel the inner body now?". Any moment you bring your attention to your breathing, to your surroundings and simply accept this moment as it is, you become a little more rooted within the peace of the present moment. Like drops of water that slowly fill a bucket, consciousness shifts from the heaviness of time to the spaciousness of timelessness. The body literally continues to feel lighter as life moves forward. There is no going backward in this shift. Short-term cycles of unconsciousness will come and go, but the deepening of awareness only moves forward deeper. Allowing true joy in life to emerge into this world. You are the present moment, experiencing itself as a human.

Evolving Exercise

Most of the physical activities within our daily routine are so familiar and repetitive to us that it can be easy to fall into autopilot a lot of the time. Falling into thought patterns that are even more familiarized and more repetitive. We still allow the activities in front of us to have some of our attention, at least enough to get things done, but never take away the attention that the mind demands for its continuous addiction to overthinking.

A great way to bring more presence into these moments is to practice being with an activity more fully. Instead of trying to get an activity over with, practice being with the one step you are on now. Practice being more alert by placing your full attention towards every action. Simply notice when thoughts come back in, as they will. Practice bringing your full attention back to the more important activity within this moment. Not without the thoughts that are needed in order to complete a physical task. Simply notice what thoughts pretend to be important when they really are not. Many physical tasks require less thinking than the mind wants to realize. Any chance the mind can get to absorb your attention has the potential to pull you back into a long narrative that has little or nothing to do with what is here now.

There is no need to grade or judge your performance in this exercise. This is just more fuel for overthinking. Failures will occur, but failures will also help you succeed and become more aware. Simply notice the presence that is already within you at this moment by bringing awareness into your breathing, into the inner spaciousness that has no narrative. This allows the deeper dimension that is always within you to become more noticeable and more easily expressed in joyful ways through the activities that occur within this moment. There is no need to strain or stress in order to bring this inner stillness into present actions. It is always within you, always noticeable when you bring your attention back into the now.

Simply practice allowing the situations within this moment to exist as they are. Practice noticing when thoughts and emotions need to resist what is within this moment. Every time the mind resists a current situation, the opportunity to become more aware of these patterns arises. The more you notice, the mind's need to resist in order to absorb your attention into negative thinking. The easier it becomes to bring attention back into the simplicity of this one step within the current activity, as every activity is always one step at a time. These are not the hundreds of time based steps that the mind must obsess over.

Being present while engaged in activities can become like an anchor that holds your attention within this one moment. The more you practice this, the less the mind can convince you that mental or emotional stress is necessary to complete physical activities. The less the mind can convince itself that it needs to just get it over with.

This allows you to bring the simple joy of your deeper existence or presence into the actions. As you become more aware that joy is not found within the doing, but is always found within your awareness of Being, you bring joy into the doing. You will see more and more firsthand how it is your inner light that brings light into this world.

Growing Ego, Growing Awareness

"Man is made by his beliefs. As he believes, so he is!"

-Krishna; The Bhagavad Gita

Within a much earlier present moment, many species of apes in Africa who were able to avoid large predators by staying high up in the trees within a vast, thick forest. Started finding this environment turning increasingly into savanna, or grassland. Global weather patterns were changing due to the South American continent's collision with Central America roughly around 15 million years ago. However, scientists are not yet sure of this exact date. This event changed weather patterns in Northern Africa, Europe, and northern parts of the Atlantic Ocean, as they were cut off from the Pacific Ocean's weather movements. Slowly but steadily turning, Europe, and more so Northern Africa into much drier environments.

The deforestation in Africa is said to have lasted till about 2 million years ago. The earliest known hominins currently date back to around roughly 3 million years ago, during a period when forests would have truly been becoming scarce within this region. This is one theory out of many as to why our ancestors became bipedal as well as predatorial. Yet regardless of the accuracy of this one theory, it simply points towards at least one of the needs that natural selection had created for our ancestors to change behaviors and evolve. As forests declined, the need to outsmart large predators increased. This situation forced a collective number of conscious beings into a steadily increasing need to become more conscious than they currently were. If they wanted to survive, they needed to become more knowledgeable about their environment, as well as more conscious of themselves.

At a later present moment, a number of different hominin species competed with each other over food and resources as they roamed the savanna. Making tools for hunting their food, as well as cooking with fire. Unknowingly growing their brain sizes by doing so. Once their intellectual abilities reached a certain threshold, a handful of different hominin species were able to follow and hunt herds of large mammals through any environment on Earth. Successfully spreading to every continent, perhaps with the exception of Antarctica. Yet, there is much of our prehistory that we do not know, and it could be possible that some hominins once even occupied parts of Antarctica.

Later still, within another series of present moments, Homo Sapiens became the most successful, as well as the only hominin species on Earth. A process of natural selection through conscious decision-making within the living beings who were involved. Creating a new level of consciousness and new ways of living on planet Earth through the ability of complex thinking. Paradoxically, at the same time, creating a later challenge for our species. The challenge is becoming conscious enough to have a complex mind without also being fully absorbed by it.

The existence of ego has been a slow but steady process of inflation on Earth that likely has its roots within a variety of many other species. It is hard to say just what level of ego we might find within other random species, but it would most likely be very different from the extremities within modern humans being lost in identification with our minds today. It could also be a phenomenon that comes and goes in other animals without taking them over entirely or at all times. Some have suggested that the ego is simply an acceleration of the fight or flight survival reaction that is within all animals. Tribalistic behaviors are found within many different species. The only real difference is the complex human mind's extreme identification with form. The form identity is often the most important part of the mind's sense of self. Creating its inability to let go of the form or forms it identifies with, both personal and cultural. Lost in the need to defend a conceptual identity as if it is more than a pattern of thoughts. Ideologies are often defended as if the accuracy of a belief structure is a life-or-death situation. To the extent that many situations can literally become life-or-death circumstances for the sake of defending a conceptual ideology.

So far, throughout Earth's history, staying alive has always been a game of survival of the fittest to some degree or another. For example, if you come across a Bear in the wilderness, there's no way to explain to the Bear that you mean no harm and no way of negotiating its response. All you can do is react quickly, hopefully as calmly as you can, and you may or may not be safe. If an invading army is storming into your town/city, all you can do is fight, take cover, or run. If you want to survive at all as a life form on Earth, you must ingest nutrients from other life forms of some kind, either plant, fungus, or animal. Not to mention having some source of fresh drinking water. These all have been the reason for many battles and wars over resources throughout human history. While at the same time helping the ego strengthen itself through feeling superior towards a labeled enemy society or labeled less important people.

The longer our species has played the game of war, the more we've absorbed it into a part of our identities. On a tribal level, it was a very useful aspect of survival. Yet, as time moved on, it became less about the survival of forms and increasingly about the amplification of the ego's form identity. If there was a resource wanted, or if expanding the lines on the map was desired. The madness of ego would be ready to go to war if necessary, through some sort of declaration that it was fate of some kind. Often outweighing the wisdom of peaceful negotiations. Simply because it was wanted as a part of the identity, personal or national, in some way, mental logic will always justify its own madness if the urge is strong enough and we are not conscious enough to see this.

Strengthening the ego through playing the victim works on both sides of this equation. On one side, the mental "I" should be able to have what it wants, and war becomes the most likely answer. On the other hand, playing the victim when the army that is lost in madness is invading and enforcing their ways onto your people, feeling as the victim in an egoic sense strengthens egoic madness within the oppressed. To enforce oppression or to be physically oppressed can both strengthen the ego's need for this moment to be different than it is. One is never enough, and the other is always more than it should be. Both make enemies out of the "others" who are seen as the cause of the madness we unconsciously create within ourselves. These patterns can only truly be broken on the physical level when they no longer exist within the mental and emotional level. However, to be awakened in such a society is to be freed from the worst part. Once our own inner madness is lifted, life becomes too peaceful for life situations to convince us otherwise.

Unlike other animals, when the human mind is reacting to a situation. Thoughts are not able to see the difference between the situation and the mental interpretation of the situation. Unable to let go of the emotional tension that every other animal experiences for a brief moment when danger appears, but have no need to cling to once they've gone. Without an overactive mind, other animals don't have the need to repeatedly think about a dangerous situation or any situation for a long time before/afterward, seemingly recreating it over and over again. With the exception of those animals who have experienced extreme trauma. Life is lived in the now.

To the mind, any situation that is resisted enough can become amplified to the extent that it can appear life-threatening. Identification with thoughts/emotions makes the past seem somewhat alive every time the mind thinks about them. This brings back some amount of the emotional energy that we might have felt every time the thought structures repeat themselves. Making it easier for the unconscious mind to identify with these thoughts/feelings. Without being lost in thinking, other animals don't become lost in mental/emotional reactions that greatly outweigh the simplicity of the situations they encounter.

Complex thoughts, at one point, had little to none of the side effects of suffering that are challenging us today. They were something brand new to nature and extremely beneficial for our ancestors, who could now work out complex situations much more efficiently than other species could before. Of course, opposable thumbs and other physical features played a large role as well. By the time the side effects of ego started becoming noticeable, usually by a few individuals not as lost in ego, who would try to aid others through different spiritual and medicinal practices. It had become too late for most humans not to be consumed by an egoic identity to some degree, as this was the evolutionary way forward at the time.

The practices of the Shaman, Spiritual Teacher, Sage, or any other labels they have been given would still help greatly in many ways in finding deeper understandings. Until we reached a point in which ego was simply too strong within almost all humans for the importance of these roles to be understood by an ever-growing identification with form. Every culture has had different degrees of success and failure in holding onto these spiritual understandings. Yet, as time moves on, the ego's growth within us cannot help but cut most humans off from the pre-ego understandings of our spiritual existence. Gradually, through the generations, the identification with form becomes increasingly all that we know ourselves to be.

Flipping the importance of getting back to our natural roots into the importance of moving beyond identification with the mind's misperceptions of reality. Creating a stronger need to become more deeply rooted within our inner Being. Ultimately, on the deepest level, your suffering exists so that you can become more deeply aware of the essence of life than any other lifeform on this planet ever has. Creating a higher level of consciousness on planet Earth. At the same time, all the species on Earth are what make life possible. The pre-ego species simply play within an earlier conscious expression of living on this extremely beautiful planet. No more or less superior than humans. Simply a different, or less conscious expression of Being.

Consciousness wants to become more conscious through you. Consciousness wants to become more conscious, though whatever living form is ready enough to become more conscious. Being lost in mind now allows us, when we are ready, to find a depth that is beyond form. On a deeper level than any other species has experienced. Introducing the emergence of consciousness becoming conscious of itself beyond the form that is being inhabited. Not without the living form, but no longer seeing our form as all we are. As some spiritual teachings would say, the cycle of life and death is broken. Not because the form will not die, but because we have found the part of us that was never born and will never die. It is part of the form we take but not the form itself. The human is still important, but the importance of Being completes the picture, so to speak. Fear of death holds the ego together; seeing beyond death dissolves it.

Many of our ancestors have already found this deeper understanding, while many more have not been ready. However, today, we are experiencing the beginnings of a global shift through a critical threshold of our collective ego. The collective or cultural ego creates an amplification of our unconscious state. An individual ego creates a tiny amount of suffering for ourselves and others when compared with a collective need for a common ideology, as well as a common "other." Perhaps an extreme amplification of the fight or flight response. Over the millennia, extreme expressions of collective ego have risen and fallen a countless number of times. Able to hold a certain threshold of negative energy within a culture for various lengths of time. Yet,always falls out of favor when an earlier reason for an inner struggle is no longer viewed as necessary. At the same time, the mind that is lost within an expression of cultural madness will see external events as the cause of this madness. On a deeper level, the ego simply wants to not just survive but thrive as strongly as it can.

Ego is an expression of life, neither good nor bad. It is not right or wrong, no matter how much the mind gets lost within these comparisons. The ego is simply an expression of life, an expression of the overactive mind. Not only does it want to live out its expression of life, but it also needs to do so in order for humans to progress consciously. Not progressing in a way that the ego can understand. Yet it is the suffering of the ego's misunderstanding that creates the progression toward the deeper truth that the ego is always avoiding; this, too, shall pass.

Through the ego's need to resist any reality outside of its mental delusion. It continues to accelerate its side effects of suffering until we no longer identify with the level of suffering we find within ourselves. Finally, the only place to go is to get away from suffering by looking deeper within ourselves. Allowing an opening for a heightened level of inner serenity to come in. This new level of awareness is the surrendered state blossoming within us. It is simply the light of your awareness that ends the struggle of lack of awareness within.

Within our species, levels of ruthlessness towards each other come and go throughout history. However, overall, the collective ego is slowly enhancing its intensity throughout the generations. At the same time, the individual ego becomes increasingly intense throughout a person's life. As a child, most people experience a mild expression of ego, with some exceptions. Small enough that there is still an opening for a certain wonder of our surrounding world to be relevant. So as not to have a need to put a label on everything. As kids learn their names, they learn the mental labels of different things. We learn our collective identity as well as create our personal, and the mind begins identifying more and more with forms. Humans begin to have an increasingly strong identification with thought. As puberty begins, hormones accelerate the level of identification, with thoughts forming identities. We find a certain peak between the late teens and mid-twenties. A peak from the strong mental and emotional energy created by hormones and the developing mind. That being said, for the person who never becomes aware of ego, its natural course is to slowly and steadily become increasingly intense as we get older.

For the unlucky/lucky ones, identification with a suffering mind may become intense enough to spark an awakening process. For most people, this becomes its own slow and steady movement beyond ego. Yet ego is no longer ego once it is noticed for what it is. It is now within this in-between phase where it is no longer what it once was, while it continues to help dissolve what patterns are left. Only the patterns that are not yet fully made conscious will repeat themselves until they have helped you become too conscious for them to continue. Only awareness/acceptance of unconscious patterns is necessary, or simply the ability to be fully here now with what you experience within. When acceptance is able to come in, presence becomes the light that ends the darkness. You can not dissolve anyone else's ego or the collective ego. Yet as yours dissolves through awareness, this can benefit those around you who are still trapped in the mind by simply no longer contributing to their unconscious madness through your own.

You begin to display our next stage of consciousness more and more by simply being at peace within this moment. Allowing others to have a taste of what they may not yet be able to understand but can still benefit from in many ways. Once we can recognize ego within ourselves, it becomes easier to see it in other people without being pulled into our own reactive patterns. This allows the ego of others to help us see more clearly what we are still reacting towards. When they do pull us back into our own unconscious reactions, we become a little more conscious of these patterns.

Regardless of whether or not we are able to explain this, everyone benefits in some way from this practice. One day, humans will be born without a need to go through an ego process. When that time comes, it will not be some utopian future but the present moment. Not perfect, but simply no longer chaotic for the sake of unconsciously needing chaos, through unconsciously avoiding chaos.

One great example of two different levels of cultural egos interacting with or really colliding with each other in many ways. The native North/South Americans and the Europeans who seemed to feel entitled to the land that the indigenous peoples were living on. Native North Americans who lived as hunter/gatherers were just as intellectually intelligent as the Europeans who were flooding into the Americas. Yet their cultures had very little need to progress in the same ways that Europeans had. There was little reason to become increasingly egoic beyond a certain balance that could be called a healthy amount of ego. Meanwhile, the Europeans seemed to collectively need to perfect an intensely unbalanced level of conceptual identity.

Native Americans probably had little to no examples of what an intensely unbalanced ego culture would look like. They had no way of explaining why these odd people from across the ocean were acting as strangely as they were. Many natives explained how difficult it was to figure out why they always seemed to be so restless about everything. Many different tribes offered help and assistance to the settlers. Creating friendly relations for some time. Many individuals, as well as small groups of settlers, were able to maintain peaceful relations, creating strong bonds between the different cultures. Yet it was only a matter of time before the wants and believed needs of a larger collective ego coming from Europe would ruin these arrangements.

With less ego or a more balanced ego, the native cultures had the advantage of not being fully lost in thought; therefore, they were still somewhat rooted in Being, more rooted within spiritual aspects of life. Still able to find inner wisdom and guidance in ways that most Europeans had lost long before this point in time. These native cultures could be just as ruthless in the ways of warfare as the Europeans, just not as eager to go to war, with less of an inner need to be at war within themselves. Native peoples, unfortunately, had the disadvantage of not being immune to most of the diseases that had accumulated over many centuries in Europe. Diseases were now flooding into the Americas, and large numbers of the population were decreasing. It is very likely that at least several societies collapsed due to diseases like Smallpciox that we may never know about. Making it easier for the collective European ego to take over these territories.

Of course, the growing immigrant societies also had a major technological advantage. They created many inventions that helped for a better life, as well as many inventions that were destructive toward labeled enemies.

Europe had become a multicultural society that was so lost in thinking that there was very little ability to prevent the tragic taking over of the two continents with the weapons they had invented, as well as the ideologies their minds clung to. The more unconscious we become, the more violent and chaotic we can't help but become. As the Europeans steadily moved westward, an unconscious machine made up of collective individuals simply plowed through the continents with little to no understanding of the madness they were expressing towards their fellow humans.

Continuously clinging to the need to get there instead of being here now. First, leaving Europe for a better life which is an understandable action in many cases. Then, leaving these first locations in the new world for a better life. Steadily heading westward with this continuous need to leave the here and now for the mental vision of a better future elsewhere. Not to forget the many expeditions of European nations in search of gold and any other valued resource with little to no concern for the native populations. This has forced many horrific acts onto the native societies, which were simply seen as an obstacle to the progression of the collective conceptual idea of the Europeanized Americas.

All of these factors have, in turn, forced many native peoples themselves to become more egoic within a very short timeframe through extreme suffering in many ways. Resisting the madness of others will always increase the madness within ourselves. In many cases, they were given no choice but to fight madness through madness and desperation of what was being lost. The natives couldn't help but reflect the European madness in their own ways. Fighting against this unconscious machine of collective egos whose only intention was to get there instead of being more fully here. A need to complete the so-called Manifest Destiny or fully Europeanized United States. This has led to many mental illness issues for the Native Americans. Many addictions like alcoholism, as well as an overall loss of much of their culture/wisdom, are torn apart by the ignorance of us versus them.

Everyone is responsible for their actions, but how consciously can one act when the mind is dragging them in unconscious directions? Perhaps if the Europeans had not been as lost in identification with thought forms, they might have been able to find a lot more helpful knowledge and wisdom from the native cultures. On the other hand, the ego's intensity grows within our species as if it is destined to become as intense as it possibly can. The irony here is that this can only eventually create the end of ego. The one thing that remains constant is change. On a deeper level, all of our extreme expressions of suffering are for the sake of moving beyond suffering. Then suffering will have truly fulfilled its purpose. Whether we understand this or not, within the short time of an individual life, consciousness on Earth continues to evolve much like life forms do.

On the surface, tragedy is horrific, and it truly is. Yet, on the deeper levels of our existence, tragedy can be beneficial for spiritual growth. Opportunity for seeing beyond our form identity. Those who suffer the most are the most likely to shift beyond the mind's continuous suffering. As tragic as the Europeanized idea of society has been. Enforcing its way of living onto North/South Americans, as well as most of the world. The diversity of cultures and the diversity of various degrees of ego creates more ability for humans to collectively become aware of the limitations within over-personalized ideologies. Allowing more opportunities to shift beyond the level of consciousness that creates these problems for ourselves and others.

The need to not be present will always produce a dysfunctional society. When everyone needs to get to the imagined future, cling to the past, as well as cling to our form identity. No one is here now, and no real progression can occur until some imagined future moment. Then, when the future comes, it is still the present moment, and we are still avoiding the same situations. The same can be said for the need to get back to a past that can never be here again. Both are clinging to a form (personal/collective) that will always be changing. As long as the ego continues to be as extreme as it is within us, it will be the most dominant aspect of our societal structures. The madness of ego will continue to be successful on a collective scale until it can no longer be collectively successful on an individual level. It will not disappear through yet another aggressive political movement that only replaces one egoic ideology with another. The external displays of our madness created by identification with thought will dissolve only when we no longer unconsciously champion our own madness through the villainization of the madness we see in others. Instead, we come to realize that only by looking more deeply within ourselves can we move beyond the insanity that is currently within humanity.

In today's world, black and brown people are often the most oppressed simply because of the technological advantages and strong ideologies that created a need for earlier Europeans to colonize much of the world. Placing strong power structures that tend to hold most of the world within a racial us versus them mentality. Few cultures have been as oppressed as much as African Americans in the history of the United States as much as a society that is lost in ego has claimed symbolically again and again that unity for all citizens has been achieved. Which has led to many important notable progresses on these issues. America's societal structures cannot seem to help but recreate similar, if not the same, forms of oppression again and again through different physical expressions for African Americans, as well as any unwealthy citizen of any race. This has meant a reincarnation of some form of slavery every time a current version has crumbled, as change is seen as a form of death to the ego. Simple changes to the systems of society can be resisted just as strongly as a physical death.

Throughout the history of the United States, racism has always been either embraced or denied as non-existent in ways that intellectually rationalize what is strongly being avoided, as opposed to consciously facing simple truths that could create simple solutions. In some cases, the narrative has gone as far as explaining that slavery is just simply the way things are. Even arguing that it creates jobs for those who wouldn't have them. These patterns can only repeat themselves until they are made conscious enough. Slavery was made illegal in the United States through the Civil War in the 1860s. Yet this didn't create an end to the concept of superiority over the mentally labeled "other."

Segregation became illegal in 1968 through the Civil Rights Act. Yet separating people by communities, making it difficult for black Americans to find education, as well as making it difficult to find well-paying jobs, is a subject that has not yet been fully resolved. Mass Incarceration and the politics behind the "War on Drugs", which could really be called a war on race. Seems to have been one of the most efficient solutions for recreating slavery in the U.S. for anyone who has black or brown skin.

As if having strategically impoverished living conditions was not rough enough. The crack epidemic struck black communities the hardest in the 1980s and 90s, as if it was strategically placed there with the specific intention to try and exterminate those within the black community. A drug that has been proven to have no real physically different effect than the powdered form of cocaine that many white/wealthy people were consuming at the same time. The only real difference is how long a person of color could be put behind bars for being accused of using. This is not to say that all individuals within any imagined form-based side have been a part of the unconscious cultural movements of us versus them. This means that not all white people are racist, and not all people of any race ignorantly hate other races. Only the majority of individuals' level of consciousness is what makes up a cultural ego. Individuals are always at various degrees of ignorance/understanding.

Today, the Crack Epidemic is long gone, but new substances can always come along, and Mass Incarceration is still holding strong. As a way to enforce African Americans, as well as other declared non-white races, into one of America's current models of enslavement. Holding many people in the U.S. prison system for years, decades, or even life. For small amounts of illegal substances, as well as other petty crimes. Drugs that seem to have only been made illegal for the sake of locking up the conceptualized "other."

Racism is decreasing through education and exposure to diversity. Yet racism still exists within a large number of humans who are still ignorant or simply not ready to realize their own inner resistance patterns towards race, gender, and other minor physical differences. Not all egoic humans have racial issues, but until we reach a point of readiness. It is simply much easier to strengthen the ego's self-image through comparing us with them. Allowing the imagined villain to strengthen their ego through playing the victim instead of looking more deeply at our own inner discomfort.

It may not appear so with all the chaos we see in the news, but our racial tensions around the world may be reaching a turning point. Not necessarily because new laws are being passed, which is also extremely helpful. More so because of the psychological aspects of oppression, which seem to be the most important for continuing to keep an "other" down, so to speak. It can no longer work on humans of any culture/race that is shifting away from identification with the mind's psychology.

The panic that is being physiologically enforced through poverty on one end, or the resistance towards poverty on the other end, among many other societal issues. No longer works as well for reestablishing ego patterns when we understand our psychology deeply enough. This, in turn, allows these problems to turn into simpler situations than they can be when the mind is struggling with its overly personalized identity with "oppression." Not false aspects of true physical oppression. Simply overly personalized thoughts in a way that holds the conceptual identity with oppression in place. Defending it as an identity of self in ways that make it difficult to let go of inner suffering associated with these life situations.

Malcolm X understood this first step. Changing his name and declaring himself free from oppression on an intellectual level. Unfortunately, he still identified with his thinking. Creating some degree of suffering for himself while declaring freedom from the externally oppressive societal structures. Unfortunately, this also declared a need for a fully segregated Black America. He wasn't doing anything incorrectly, just that everything he did must have had various degrees of unconscious mental/emotional suffering whenever his thoughts became rooted within resistance, as opposed to creating a new form of acceptance that would aim towards uniting all Americans. Given the situations that Black Americans faced in the 1960s, it is very understandable to see things the way he did.

Martin Luther King used a peaceful march strategy to unite all Americans. Yet he also had the disadvantage of suffering through an identification with his mind. Both of these men, as well as many others, like the Bus Boycotts in Montgomery, Alabama, achieved remarkable things. Acknowledging that they also had limitations because of ego. Simply demonstrates that anyone can find themselves doing amazing things. The more we surrender to the tasks at hand, the more we can achieve within these tasks. All of these movements were swimming upstream against the currents of society, and it is amazing how much was accomplished. Yet, on the other hand, the majority of African Americans in the 1960s were ready enough for real change that nothing could stop it from occurring.

From what I understand, both Malcolm X and MLK took some degree of notes from Mahatma Gandhi's strategies with his peaceful protests against the English Empire's colonization of India, MLK more so than X. Through a method of not partaking in any violent demonstration as the English tried again and again to set up a scenario for. The English Army's effort to make the people of India look like a "violent race" backfired when they remained peaceful in their protests through his ability to have compassion for all the humans involved on both labeled sides. Gandhi was able to turn a giant problem that seemed impossible to solve, according to his peers. Into a simple situation that he helped solve in simplistic ways thanks to no longer needing to identify himself as the victim of the situation. Not without physical oppression or physical challenges. Simply no longer lost in a psychological identification that is imposed onto the situation.

This doesn't mean that anyone who helped create change through the Civil Rights Movement was incorrect or wrong in the ways they pursued justice, including those who found themselves in violent riots. All of these people took action when action was needed, creating change that benefits all Americans. Humans can only ever act through their current level of consciousness in order to create real change in societal situations that are lost within the resistance of ego. We must first understand how we ourselves become lost in unconscious resistance patterns. For the unconscious mind, there is no situation without the need to imprint the illusory identity onto it, for the continued survival of the ego. This is the very reason society creates these dysfunctional situations that become extremely difficult to move beyond. For many people, over-personalized thoughts of how a situation should be will not likely allow any space for new solutions to be discovered.

Natural selection and cultural conditioning have created a slow but steady inflation of ego for eons now. The mind's need for continuous thinking has often created feelings of something missing in life. Not quite knowing what, but feeling like something is lacking in the conceptual version of "my life." Often creating the need to fill a supposed void with something new or a need to get back to something from the past. When satisfaction is found, which may not be very often, it doesn't stay for long until the feelings of being incomplete take the mind over again. People may try to fill the void through their work until work becomes the conceptual grind, which is yet another problem created out of simple situations. They may feel a need to always chase after that next promotion or some alternate position that must be better than where they are now, unknowingly always running from themselves. It can be very beneficial to climb up the ranks at work until the thoughts about getting from here to there create too much stress to be fully here within the current position now.

Many people will decide that dating, getting married, and maybe having kids will fill the so-called void that the mind views as a problem. This can be very helpful for living a richly fulfilling life in many ways. Yet, at some point, the mind's inability to be satisfied might begin to overlook how much wealth they've found through family. Simply because the thoughts they identify with still are unfulfilled. The mind cannot understand that the so-called void that is trying to be filled/avoided is our deeper self waiting to be recognized if/when we become ready to let ourselves be as we are.

The overthinking mind is only satisfied for so long before some new reason shows up for why it is unsatisfied. Some people might get divorced in this scenario or cheat on their spouse. Some people might look for new activities/forms in order to try and feel fulfilled. Maybe a new hobby, which there is nothing wrong with, will at least distract us from this unexplainable emptiness that we feel inside. The emptiness that simply looks like it needs to be filled because it is not made of form. The only conceptual role this inner void fulfills for the form identity is to be covered up for the sake of strengthening the identity. No formlessness can be acceptable for a strong identification with form to feel good about itself.

Maybe simply buying new things can bring in some kind of satisfaction through forms. Nothing wrong with buying new things, but when it's fueled by ego, it will likely bring at least as much dissatisfaction as it does satisfaction. In many cases, going out shopping can feel very satisfying for a brief time as we bring the brand new shopping bags home and feel renewed in some way with our new merchandise. Until the mind feels these things aren't so new anymore, which can be a very short process. Seemingly out of nowhere, the same old feelings of being incomplete are all the mind wants to think about again. Dwelling within a personalized narrative of why life is never quite as it should be. Sometimes, a product lasts for a long time and can be very fulfilling as long as it is around. The only catch is the investment of needing this form to bring us fulfillment. Our self-esteem becomes reliant on the external entity to bring us satisfaction. When these forms no longer function as they once did, the ego feels a certain kind of death has occurred. This can be a perfect time to become more self-aware.

Maybe buying a whole new wardrobe can do it. Maybe reinventing the mental persona through new clothes can help us feel complete for a short time. If it can be afforded, we might go out and buy a new car or even a new house to live in. Suddenly feeling, renewed in a deeply needed way. Until the same patterns of feeling incomplete come back around yet again. No matter where we go or what we do, this feeling of incompleteness always seems to be with us in some way. Yet the mind is always sure it can find something to cover it up with.

The appearance of the body can be the largest out of all the ego's obsessions with form; after all, it is our physical form. Yet, we are the only species to worry about our looks in the ways that we do. Dogs or cats, for example, do not worry about the body's physical appearance to any degree that humans do. It simply is as it is. Like most species on Earth, they are below the ego level of consciousness. For them, form identity is not a thing to chase after. There is individuality, but it's not clung to in the ways that the overthinking mind cannot help but need to. For the egoic mind, the body must be perfect to some degree, or else it cannot feel whole or good enough. Body dysmorphia is not something that only extremely depressed people suffer from. Most people experience some level of dysfunctional views of their physical appearance.

Positive and/or negative versions of seeing what we think we will look like in the mirror become a main factor of our egoic persona. How we look at ourselves becomes a definitive objective, always needing to look a certain way. Nothing wrong with the way we want to look, physically, as well as the clothes we choose to wear. The mental identification with how we look is the only difference between a healthy view and a dysfunctional one. There is a need to look a certain way, and in many cases, there is need to never age physically. It is the restless chasing after what we unconsciously want to be perfect that strengthens ego. Not how we look but how we should look. When the body is allowed to be as it is, the form identity is dissolved. This may not fully occur instantly but will occur as you become more ready. The first step is to understand the ego for what it is. Simply noticing your relationship with your physical form through ego is the beginning of the end for the ego.

In the same way, wolves will nip at each other with their teeth, maintaining social order and keeping everyone in line for their hunting strategies. Human egos thrive on keeping the personal/cultural ego in place by nipping at each other in many different mental, emotional, and physical ways. Wolves most definitely experience many mental/emotional aspects within their daily lives. It's also very likely that within species like wolves, who have complex minds, there is an egoic quality to some degree. The biggest difference is that, unlike humans, they are never fully taken over by the dimension of thought.

This effect of awareness being absorbed by the mind comes and goes in various degrees within many different species that are still within the pre-ego level of consciousness. Still rooted in Being, but not in a way that can be fully recognized by the individual. These animals benefit by being rooted in a deeper wisdom in many ways but without a deeper knowing of self or the realization of our connection to all. This requires an Ego Stage for a species before moving to the next level of consciousness, that is the one consciousness becoming consciously aware of itself, beyond the form, but through the form. Finding infinite peace in knowing ourselves more deeply. Species that are below the level of ego are unable to be completely absorbed by endless thinking in the way that humans currently are. Perhaps it is a lack of a frontal cortex like human brains have. Whatever the physical aspects, humans have been long since fully absorbed into the dimension of thought. The Human equivalency of ego has evolved to the point of not just absorbing all of our attention but thriving on keeping it this way.

Sometimes, when people strengthen each other's ego, like wolves nipping at each other, it's a friendly back and forth of some kind. Other times, it can get very serious, as well as anywhere in between. Anytime that the interaction enhances the mental identity, there is some sort of rush of energy for the ego to absorb. Nothing wrong with this; in fact, the more noticeable it becomes, the more we move beyond the ego level of socializing. Pleasantries and repeating familiar thoughts back and forth can be helpful for holding personal identity together. However, anything that feels bitter, for the lack of a better word, seems to strengthen ego the most. To say anything that is negative wouldn't be quite right. As negative is never without some amount of positive. Resisting what is within this moment is what really strengthens a mentally defined separate entity. The story of "me" that is always time-based. Our identity can be strengthened by our positive experiences as well. Yet it's the negative-based or bitter feeling identification aspects that seem to hold more intensely. Through the fear of losing our identity or not knowing who we are, mentally speaking. The self-image must be maintained.

This is the fear that allows the ego to gain more strength for itself within negative mindsets. Add hundreds of thousands of years of this continuous egoic behavior, and you might get a world as dysfunctional as we live in today.

The need to defend an idea, or viewpoint as if it is who we are. It is simply because thoughts can only understand thoughts. Holding opinions makes the ego feel important in a way that holds the thought identity together. A current collective opinion that is agreed/disagreed with amplifies this mental sense of self. The need to judge others has more to do with how fragile a mental self-image feels about itself than it has to do with what the mind thinks or says about someone else and someone else's ideas. When we are content with ourselves, there is no need to criticize anyone else or whatever ideas they hold. It becomes easier to feel compassionate for others, whatever situation they might be in.

When the ego displays a well cleverly crafted expression of criticism towards another, a false sense of satisfaction or a false sense of self is inflated momentarily. Until this momentary feeling fades, and the dissatisfaction being covered up creates the opposite effect. Building a need to continuously criticize more people and situations for the satisfaction of feeling superior towards someone or something for another short moment of inflated ego. Creating more inner resistance within ourselves as this need continues.

Criticism affects everyone differently. Some feel more like themselves through criticizing. Other people may feel more like themselves when being criticized, as well as anywhere in between. The reactive patterns may be different, but it is the ego's need to take life/themselves too seriously through over-personalizing situations that keep the repetitive patterns moving along this path. Until if/when we begin seeing them more clearly for what they are. Ego holding its illusion together.

The need to prove a point or argue an opinion always has more to do with defending a mental self-image than it has to do with the subject at hand. Someone can understandably feel extremely compassionate or concerned about a specific topic within politics or societal structures. Despite this, as long as the ego is involved, an intellectual conversation is easily reduced to what might look more like a grade school brawl than a mature debate. Using name-calling in some cases. As each side is more interested in defining the other side as moronic, then they are interested in having a truly intellectual conversation that investigates truths that may have been overlooked. Defending the possibility of being "wrong" one of the ego's greatest fears. This means that a time of realizing we are wrong about something is a great opportunity to move beyond our egos. Because simply seeing it for what it is brings the light of our awareness into the darkness of its illusion.

Who will we be if the ideas we cling to are not absolutely right? This feels incredibly embarrassing to the ego and creates a need to defend an idea as opposed to learning new information outside of its identity's current perspective. As long as the mind knows it's right, a certain level of worthiness can be felt. A kind of distraction from feeling incomplete. To be wrong is like a death to the ego in a way. Now, the deep void that is not of form can feel more devastating than before, when at least the feeling of being right could temporarily cover it up. Instead of investigating this inner emptiness, the ego would rather feel the temporary relief of conceptually reestablishing the mental identity. Repeatedly reminding ourselves that the mental image is who we are. For thoughts, the image of ourselves seems very clearly to be all that we can be.

The ego resists looking into the void because it doesn't want to wake up. You could say that it resists seeing beyond its dream, beyond its obsession with self-image. Yet once we recognize the patterns within ourselves, we do so from a place of inner emptiness, a deeper awareness than the mind. We've already begun to awaken. Ego doesn't know that it is ego. Recognizing it dissolves the identity within it. Beginning the transformation of resistance into acceptance. No level of inner resistance can survive your full acceptance of it. It may require some practice, but inner presence/acceptance is all that's needed to awaken out of ego. A new level of consciousness is emerging from the old.

The ego may find temporary relief from maintaining the individual identity through becoming a part of a collective identity. Sometimes, in religious organizations, maybe within certain employment agencies, social justice causes military, government, or following political issues, and so on. This can seem like an escape from our personal suffering. Yet when the group is an accumulation of unconscious individuals, the opposite may occur. Now, we may find something less conscious than within our individual suffering. We have entered a collective amplification of suffering. The need to change a situation, or situations for a good reason, can be a great thing. However, mixing this with an ego's need to declare and strengthen the victim role will cause the level of unconsciousness within the individuals to become much lower and stronger than it was before. Suffering has been amplified many times, but now, it is the collective ego that must be defended at all costs. Instead of continuing to look for what is best for all within any given situation, now the predetermined ideology is the only answer for any situation.

One of the most extreme expressions of collective unconsciousness is when it is amplified through a need to engage in warfare of any level. An exercise that has shaped ego for at least tens of thousands of years. Becoming less and less purposeful as a physical survival strategy and more purposeful as an ego survival strategy as the present moment moves on. Whether it's the need to invade, a need to defend from an invading force, or simply being within this madness without any personal contribution towards the violence, the extreme energy of suffering that war creates is experienced by all who are within any region that is affected by the combat as well as those outside of the battlefields, who either support their identified side or mentally/emotionally resist the conflict. All unconscious minds of those who are physically and mentally involved will experience some amount of strengthening of ego during these situations. On a deep enough level, there are no sides. War creates suffering for the victors as well as the declared losers.

The invading force doesn't necessarily realize that they are also creating as much suffering for themselves. If they do, it can usually be intellectually worked out as a means to an end if the mind is invested enough in the cause that is believed in. The rush that is felt within the violent actions, can give off certain feelings of being more alive than in any other life situation. Seeming like a positive rush toward the goal of victory. This can also be experienced on a smaller scale. When people's minds become absorbed by political views that are rooted in resistance. Collectively, their level of consciousness has the potential of dipping toward the direction of denser, less conscious ego expressions. The collective narrative fuels a need to express inner resistance in more and more extreme ways. Clouding the truth that what is seen as a positive rush is actually a rush of inner suffering.

Within any level of warfare, whether or not the experience has been confused as positive. The mental and emotional negativity has consumed them in a way that they are no longer themselves. They are like a possessed version of themselves that is now enjoying the negative energy in a way that will continue to seek as much madness as it can for as long as it can create the same rush of negative energy for the sake of maintaining this heightened level of ego. The extreme negative emotional ego is who they've become, for now. Then, after the conflict is over, collectively and individually, they might simply snap out of it and become the normal version of their egos that they had experienced before. However, this will very likely be altered by their experience in certain ways. They are also likely to have some level of trauma that will stay with them if they cannot become conscious enough of its source within. These patterns will sometimes lie dormant until certain things trigger extreme emotional negativity again. Most commonly known as Post Traumatic Stress Disorder.

Unless they become conscious enough of these patterns, they will likely repeat themselves in some way throughout their lives. They may even create an unconscious need for short bursts of violence in some form from time to time, inflicted on themselves and maybe others. In some cases, only verbally, but if patterns are strong enough, they can become physically violent as well. The more unconscious we are, the more violent we cannot help to become. These expressions of extreme suffering will last for short periods of time until the dense emotional ego has been fed and goes back into its dormant state, where it stays until it is ready to repeat these patterns again. This type of trauma is only one of the many ways that a heavy emotional ego can become active within us. This subject will be covered more deeply, with much more detail, in a later chapter.

In different places, at different times throughout history, many humans have created extremely horrific acts onto others through a strong belief that they were doing what God wanted them to do, as well as for many other reasons. Not because they were stupid, weak-minded, or even evil in some way. Simply because their overthinking minds become increasingly absorbed by a collective narrative that expresses a need to rid the world of what they believe to be the cause of suffering within themselves. In many cases, the mind's fear of the unknown can create a problem that doesn't exist.

Through expressions of collective ego within these scenarios, a sort of warfare culture has built up and has been maintained within our species for thousands of years now. Wars come and go, but the same state of mind stays as the patterns continue to reestablish themselves. Times of peace cannot last for too long when the collective individuals are always fighting battles within themselves. The need to create a defense from invading forces can be a very wise decision. Unfortunately, given enough time, the right political, monetary, ideological, or just about any reason will create enough of a need to display just how strong these military forces are. Until we finally collectively become conscious enough, conditioned societal patterns will repeat themselves, and wars will come and go in the same ways that an individual's patterns of stress will come and go. In some cases, all that is needed for military conflict is a General or top official who is already war-bound before they take their political and or military position.

For the unconscious mind, war can appear as an art form that overlooks the extreme madness involved. It can be seen as a sort of chess game of violence. Madness for the sake of strengthening and displaying the strength of madness in many ways. Becoming the ego living itself out in one of the most extreme expressions that it can. Ego defends a conceptual form by destroying other conceptual forms for the sake of winning a battle of egos, as well as expanding its conceptual form. The conceptual battles are always a little more important than the physical reality. Ego becomes more important than the livelihood of the people involved.

One cannot strike down a supposed enemy without creating a need for retaliation. This wisdom becomes lost within the mind's need to achieve victory in some conceptual way, even if human lives and valuable resources are at stake. This, too, becomes more conceptual as the mind becomes less conscious in its reaction-based thinking. There would be no World War 2 without the First World War. No Cold War without the two world wars. There is also no saying what is to come in the future, as the Cold War didn't resolve relations in a conscious enough way to end the basic idea of us and them within our global politics. To have different countries is one thing. To need patriotic enemies is madness that will create more madness until these issues are fully resolved on a conscious enough level.

When the mind has become the character of the soldier or officer, the costume that is called a uniform accelerates the egoic image that the mind is lost within. This accelerates the need to play war or act as the character who is playing war. Aiding in the suppression of many emotions of trauma that are experienced. Simply because the thoughts have a clearly logical reasoning for why they are not important now, the only moment there ever is. More important now is winning the game of war. A strong strategic move against the foe, or vice versa, will excite the mind in a way that overlooks the madness involved. The mind has worked out its reasoning of why the cause is more important than the overall effects. So now that compassionate concerns are out of the way, the ego has a chance to shine in more ways than it could before. Creating more and more normalcy within these scenarios.

The more time we spend within normalized patterns of insanity, the more insanity becomes seen as necessary. Over the millennia, the cultural normalcy of these patterns can only seem to accelerate and strengthen their supposed need. Even a worrisome threat of war can be enough to feed these patterns within a society. Sometimes, this can be enough food for the collective ego patterns as they rise and fall in and out of dormant and active phases within us. We feed and sometimes strengthen our collective ego, which we are all connected to on a deep enough level. Other times, within a culture that is already suffering enough through collective inner resistance. There never needs to be much of a pull toward violence if we're already unconsciously looking for it. Sometimes, even the slightest believed threat can launch a culture's collective ego into action. An intense threat will almost guarantee a need for the culture's individuals to fall into the level of madness that thrives on expressions of warfare toward the supposed others.

Playing the victim is one of the best ways for the ego to strengthen itself. This works on both sides when engaging in warfare. Defending one's homeland from an outside invader may at first have little to no "victim role" involvement. Allowing a strong physical retaliation that may or may not be rooted in the concept of us versus them. Simply defending the culture from the madness that is invading. Whether or not to be involved in violent acts for the sake of keeping the invaders from consuming the resources as well as the culture is a tricky question to answer. Should they simply stay peaceful and allow the madness of the opposing army to take them over? In some cases, this is the best strategy. It can also be helpful to simply wait until the initial insanity has ended and reclaim the culture's societal structures at an appropriate time, if possible, which could take a long time, as we've seen with Colonialism.

However, if it is through relatively short acts of self-defense, the madness of only a few battles may not be enough to pull the collective into a deep state of suffering before the invaders are forced to leave. This can be a very tricky choice to make. Most wise or Enlightened people would say that peace is always the only answer. Even short-lived acts of violence can create enough suffering to affect the collective for a long time afterward.

Whether it's encountering a one-on-one act or a collective act of violence, more important than the actions or non-actions that are taken would be the inner awareness one has of one's own potential towards inner resistance during the situation. Suppose the opposing threat is seen as nothing but a horrible enemy. The choices will likely be made from a similar level of madness that created the enforcer's need for this violence if there is enough presence within the humans making these choices or within their actions. They will more easily be able to look for solutions that point towards sanity. More so than unconsciously acting in ways that accelerate their own inner madness. The ego will only see its version of right versus wrong. An acceleration of my ego that will prevail over their ego. Not that it will be seen this clearly.

Madness cannot see that it is madness. It can only see it within the "other." If we see madness within ourselves, it dissolves, and so does our need for the labeled other at different times in the past. Almost every region of the world and every conceptually defined race has owned human slaves. This is a practice that has taken time for us to realize more clearly the level of madness involved. In many ways, these patterns reinvent themselves through different arrangements of labor, living conditions, and illusory superiority over others. Any unconscious pattern will repeat itself until it is made fully conscious. It becomes a good example of how progression is slowly made as we gradually become more aware of the madness involved in earlier decisions. The one thing that is always constant is change. Because of this, planet Earth is continuously evolving.

It can be slow-moving, but in many ways, intellectual education can allow us to become much more evolved as a species. In terms of compassion for all beings, this tends to require seeing beyond the ego. Allowing us to become more compassionate of ourselves, which in turn allows us to see that there is no "other" in the sense that ego defines it, allowing us to become more compassionate towards all living beings. It requires a certain alertness in order to see beyond personalized boundaries of understanding. Even when we support the idea of unity for all, the ego is clever and can creep in if we are not alert enough. Small details are easily argued against when the mind is looking for an identity to defend. The mind can create madness as it resists the madness it is fighting against. It requires an alert presence to avoid getting pulled into these patterns. Yet, as we become more aware, madness cannot survive our direct awareness of it within ourselves. It begins to give itself away more often the longer we stay vigilant, looking for these patterns. As long as we identify with emotional resistance patterns, they can live out as if they are a part of us. Until if/when we can see directly that these patterns are not who we are. They are simply pain that has accumulated from the past, unconsciously being reexpressed through the ego today.

In the case of the collective ego, at this point, we are all born with various degrees of how our ancestors expressed their egos in the past. If this is looked at as a burden for us to carry, this is more ego. Suppose this can be understood as the reason why we are now awakening collectively from our species' past suffering. Then, it has fulfilled its purpose within you, and you are moving beyond what those before us were not yet able to understand. As we awaken, we honor them by moving beyond the suffering which they were trapped within. You may even become grateful at some point that they have helped you reach this place within our evolution. How your parents and other family members acted was only through the level of consciousness that they could understand. We can hold onto resentment and hold onto ego, or we can make these patterns more present within ourselves. Creating the change our world has been waiting for.

Out of all the collective ego expressions of war that can be felt within our species in different ways today. World War 2 might be the moment of extreme suffering that cracked our global collective ego shell enough to begin our current expression of collective awakening. Humans within Germany, as well as Japan, found themselves pulled into such an extremely unconscious state that the violence that was carried out seemed purposeful to them. There is no greater evil than the mental labels the mind places on the "other." As well as there is no better motivator for violence than inner suffering. It is also very likely that the invention of atomic weapons and their potential for extreme destruction has a large part to play in our collective need for spiritual depth and wisdom within our current global ego state.

Added to this could be violent movies, TV shows, and video games that our egos can now live through vicariously. Strengthening our inner suffering reactions can be especially strong within the powerful effects social media can have when we take technology too personally. Technology itself is neutral until the ego needs to make things extremely personal. All of these elements contribute to the earliest stages of this shift in our consciousness, the critical threshold of ego. Without a critical threshold of suffering, there would be no reason to move beyond the ego stage.

This is why there are so many different expressions of spirituality throughout human history. We are always looking, with various results, for what we know is here, but we cannot seem to see it. In many cases, it can be easy for the mind to become extremely fascinated with the intellectual aspects of religion and or spirituality. Be that as it may, the only reason we find ourselves seeking answers on this subject is because a deeper part of us knows that something is missing within our mental identity. The mind avoids looking more deeply than its identity, but the depth is here. The mind chases after external elements, becoming lost in complexity and spiritual form identities, while what we seek is always deeper within us. The spiritual ego does not know that it's in its own way until we can see more deeply how it gets in the way. Any religious/spiritual ritual that is rooted in wisdom can work to enhance ego, as well as point us deeper. It all depends on how ready we are to move beyond our mental identity.

While the mind is a valuable part of the human experience, the mind is also only one organ within the body. One that can be used much more efficiently when we know ourselves deeply enough. As the saying goes, deep roots don't fear the wind, and it becomes difficult for the mind's story of time to pull us away from the now when we are deeply rooted in Being the one consciousness, what the ego is unconsciously avoiding is the complete annihilation of its inner suffering that is identified with. Being fully accepting of this one and only moment of life.

The percentage of individuals who are awakening out of the conditioned mind is still low, and ego is both stubborn and very clever. Yet growing technologies are helping us connect in many new ways, helping us see the global threshold of ego. Strong evidence that we are ready to begin shifting as a species. As suffering is necessary for the need to become more conscious. No mud, no lotus, as the saying in Zen points towards. The depth that we find within is what dissolves the conditioned mind. As more individuals find this deeper truth within, the more they accelerate this collective shift within us all. This is not through a new belief structure or a set of moral values that thinking can turn into hypocrisy to some degree and recreating ego, but more deeply through our collective Being, becoming rooted within our collective spiritual dimension. Where wisdom becomes more important than opinions. Peace becomes more valued over chaotic reactions. Stressful/dense frequencies of consciousness begin to more easily collectively shift into higher/peaceful frequencies of the next level of consciousness on planet Earth.

Becoming more aware of the present moment is the only goal for freedom from the conditioned mind. Moving beyond suffering is not moving beyond challenges or painful experiences. It is simply moving beyond the resistance of these things. Moving beyond identification with the resistance towards these things. Society can be like a giant ship that takes some time to turn towards a new direction. However, as more people practice becoming aware of themselves now. More naturally, a dysfunctional society will balance itself through individuals, within collective individuals, as the two are paradoxically one and the same. You and many other people are a part of this global shift that is occurring now.

The more we can accept what is within this one moment, the only moment life ever consists of. The more peace we realize within ourselves, the more we bring this light into our surroundings, and the more we will see this light within others, regardless of how conscious they might be of themselves. Their light is there, just covered up by overthinking. It can be seen by those who know what they are seeing. You are the light that our world has been waiting for. In the book of Thomas, one of the Gospels cut out of the orthodox movement in Christianity, Jesus says, "The kingdom of heaven is within you, and one who knows thyself will find it." He was not pointing towards a future heaven that can only be found after life, but the heaven that is within you now, life itself.

One day, people will look back at humanity's horrific actions while knowing deeply within themselves that no one can act outside of their own level of consciousness. Understanding that in order for life to continue evolving, there must be intense challenges sometimes in order to create the need for the next step in evolution. As more humans become aware of their overthinking, life on our planet will naturally become less chaotic as we begin to find more joy within ourselves, more joy within the present moment. Allowing this joy to be increasingly found within the world we are a part of.

If the mind's dream is a pleasant one, there's little to no reason to wake from the dream. When the dream becomes a nightmare, however, the readiness to awaken becomes strong. Humanity is at the end of this long-lasting nightmare, and we are waking up on a collective scale. Yet the more of us that awaken, the easier it becomes for us all. Soon, we will come to a present moment where the general public is consciously aware of the personal and collective ego, allowing this transition to occur more easily on a collective level and creating the next step in consciousness on planet Earth. Your awakening is just as important as everyone else's. A new world is beginning with you within this present moment.

EVOLVING EXERCISE

Life is never quite as serious as the mind often needs it to be. One of the simplest exercises of becoming more watchful of habitual mental patterns, is noticing when the mind needs to take this moment too seriously. When it is unable to notice the joyful simplicity that can always be found within the background, or on a deeper level, within any situation. Usually by over personalizing what we experience. There's nothing wrong with being yourself in any situation. The issue with ego is that it's a pretend self. We are never without ourselves. Yet when we take things too personal, it's not the truthful "me" who is doing so. It's the story of time being reimposed onto us through resistance towards what is.

A simple exercise of noticing when you are reacting through ego, as opposed to simply being yourself. Is to practice accepting whatever may come within this one moment. Not saying yes to everything, just simply noticing the difference between what simply is, and resistance towards what already is within this moment. All that is needed is an alert presence, and a willingness to accept what already is, to the current extent that is possible. There is no right or wrong, any noticeable failure so to speak, is a success in knowing yourself more deeply. Only the ego is resisting, to resist that is simply more ego. To simply notice that, is moving beyond ego, one present moment at a time.

When life is no longer taken too personally, a certain joy emerges, allowing all of life to be seen as joyful. Even in the more stressful situations, joy can still be found within the background when we become rooted in our Being. As you continue to dissolve the mind's need for resisting what is, the Heaven that has always been within you can now be experienced as the simple joy of life itself. Equally through every situation and life form that you encounter. By simply accepting fully what is within this one moment. Situations that could be viewed as a tedious experience for the mind, are transformed into this one simple moment of freedom from identification of the mind. As life becomes too joyful to take things too seriously.

GOING INTO NATURE

"What was said to the rose that made it open, was said to me, here in my chest."

-Jalal Ad-Din Muhammad Rumi

13th Century Sufi/Muslim Poet

Sufism is a religion that began long before Islam but merged with Islamic rituals in order to still be useful for those who seek a deeper understanding of themselves. As well as a deeper understanding of their Muslim faith beyond the ideological interpretations of the religion. Instead of becoming forgotten within a rising Muslim culture, Sufism adapted in order to continue its teaching. While also utilizing Islamic rituals as a way to look more deeply within. Displaying how most religions can be rooted in wisdom if those who are participating are conscious enough to be rooted within their own.

Fita is an Arabic word used in Sufism, meaning one's own true nature. In order to think about these arrangements of letters, the mind will decode the patterns and maybe go into the past history of remembering these patterns. Perhaps remembering when we first learned about the word Fita, or perhaps beginning a web search on the history of Sufism before/within Islam. Which can certainly have its place. Maybe the mind will become satisfied for a short time with this new or renewed knowledge until the next thing seems to need the mind's attention.

Words are very limited when trying to point beyond them. However, a word like Fita, or presence, can be used as a reminder to be in this moment more fully than the momentum of thinking that is pulling us away from this one moment. We can know, on a much deeper level, what words like Fita are pointing towards. Being present is an art. It is a practice that creates more and more inner peace awareness within our lives the more we practice. It is an art because every moment presents a new potentiality for you to be here in the now more fully than the mind is used to being or may want to be. While this practice is always the same principle, your expression of being here now is unique to anyone else's. Simply because you are the light that is conscious of the world you are within now. You are the unique conscious observer.

When the mind looks at a plant, it sees the name it has been given and probably a few other intellectually based details about it, like the shape and color of its leaves or the shape and color of its flowers. Maybe its medicinal properties, which certainly have their place, but the mind is unable to see the depth that any plant or animal can show us when we take a moment to simply become still. Simply become present with the expression of life or Being that is within the form. Reflecting a deeper sense of oneness within ourselves.

Some Zen Buddhist monks will often meditate with flowers in front of them in order to help find awareness of Zen within themselves. As a flower opens itself up to the world in order to spread/share the plant's genetic material, it displays a kind of serenity. It is the space that brings out the flower's beauty. Similar to the journey of our shifting consciousness. A flower requires time to be in a physically dense state as a bulb in order to transform into a delicate display that can attract insects and other animals towards its beauty. Much like within an awakening human, it is this dense state of the bulb that creates the space within, allowing the delicate flower to shine. It is the intense density of inner suffering within us that allows spacious awareness or inner peace to emerge beyond the ego state of consciousness.

Zen Buddhism is said to have originated in India within the 5th century BCE when Shakyamuni Buddha realized awakening while in meditation. This form of Buddhism, which focuses mostly on "Zazen Meditation" (sitting meditation), was practiced only in India for about 1,000 years before it began to spread throughout most of Asia. The term "Zen" is derived from the Japanese pronunciation of the Chinese word "Ch'an," which is the Chinese translation of the Sanskrit word "Dhyana," which simply means meditation. At the same time, Zazen can be the primary focus of Zen in the monastery traditions. Other everyday rituals are utilized as well. Simple tasks like working in the garden, as well as completing chores around the monastery, while practicing the art of doing one thing at a time. Often being reminded by Zen Masters to be more fully with the tasks, and less in the mind.

Some Zen Monasteries will send monks out to work in nature on a regular basis in order to gather materials like drinking water, and firewood. On the surface, these rituals can seem to only have a practical purpose for the Monastery. However, as the practice of Zen continues, the monks begin to find that these tasks are also designed to help them find the essence of Zen within themselves. The old phrase, "Before Zen, chop wood, carry water. After Zen, chop wood, carry water". Refers to the two different points of view as a Zen practitioner goes into nature everyday in order to complete these tasks. When the momentum of thinking is still strong, going into nature can be pleasant for a short time until thoughts start taking over their attention. Before long, the beautiful forest experience is replaced by thoughts about how things should be this way or shouldn't be that way. Thoughts like, "Why do we have to walk so far into the woods!?" "So much water to carry, we should hire someone else to do this!" "Why are we always chopping wood? Can't we do something else!?" "We have very little time to truly get to enjoy this forest, all we ever do is work, work, work!"

As the awareness of Zen builds within the individual, it becomes more clear that the only problems that the monks are having are in the mind. Suddenly, the work becomes secondary to where they are now. Without the mind's story of time dominating the experience, every moment becomes simply one thing at a time, and the present moment becomes timeless. All the life forms surrounding them help with this experience. As the wilderness is no longer seen as many separate, conceptually lifeless entities. Instead, their living energies become present, and they can simply be seen as the universe expressing itself as a living essence through every life form. Different forms, but the same one consciousness that is now experienced more consciously within the awakening individual. The seemingly arduous tasks become celebrated instead of resisted. Life itself becomes more of a playful dance as opposed to the struggle that comes from defensively needing to get away from this moment. The present moment becomes infinite as they continue to perform the tasks one step, one present moment at a time.

This is one method of going into nature to become more deeply aware, but it is not for everyone. Only you can decide what is right for you along this path. Maybe going into nature every so often to hike along a trail or have a picnic is a better way to connect with the wilderness than the everyday physical labor Zen monks might experience. Whatever methods you find to be helpful for you, going into nature from time to time is highly recommended for spiritual growth and deeper understanding.

If you live in the city or suburbs, as most people do, the amount of wilderness is extremely limited and well-domesticated. Leaving the human-made structures behind and journeying into nature can always remind us that humanity and our thinking minds are only one expression of life on this beautiful planet. If we approach the vast amount of life forms with a certain amount of inner stillness, the wilderness will reflect that same stillness, allowing it to become more easily noticeable within. They are expressing the same stillness, the same one consciousness being expressed through every living form in each one's unique way. This truth is not experienced through an intellectual knowing but understood from a much deeper place within. This is not to say that all other species will be peaceful with us if we are peaceful with them. If we encounter a large predator, for example, in the wilderness that can be dangerous to us, we must be aware of how they may act towards us. Being present can be very helpful with this as well. Yet, much like within a lot of humans, you can notice the essence of their Being when you know it within yourself, but proceed with caution whenever necessary.

Whether the mind is overthinking or not, the stillness of the trees and other plants in a forest can reflect a depth within you that the mind cannot conceptualize and will likely prefer its own thinking to not thinking. Yet the more familiar we become with this depth, the less the mind can convince us that the repetitive thoughts are more important than becoming more grounded within the essence of life within ourselves, especially in situations where we are surrounded by life that reflects Being back towards us. The simplicity of nature helps point us towards the simplicity of being without thought and the joyful feeling that comes with this.

Many religious stories, whether they are true, allegory, or completely mythology, have served as a source of wisdom to various degrees throughout human history. Depending on the level of inner depth found within those writings, as well as interpreting them. Most religions talk about finding God through nature in some way or another. In the book of Exodus, Moses is said to have gone into the desert and climbed Mount Herob, an unknown mountain today, with an unknown location. He was said to have found a burning bush, which God used to speak through and guide him along his path. The bush was described as being on fire but not consumed by the flame.

Moses asked what name he should give, when people asked what God's name was. He was told, "I am that I am. Tell them I Am sent you". "I Am" is a name that makes it difficult for the mind to put into conceptual form. As "I Am" is not a thing but a state of Being. A name that, when noticed, cuts through conceptual labels and the separation of conceptual forms that ego cannot help but create. Pointing towards the oneness of all beings that is within us all. The phrase "I am that I am" can also be used as a way of pointing towards surrendering to what is. No matter where we find ourselves in life, this moment is as it is, or we could say, within this moment, I am that I am. There is no deeper wisdom, no deeper peace than first fully accepting this moment as it is before we decide if an action is needed. When Moses was asked if he had seen God, he said his eyes could not look upon him. That he didn't speak out loud but revealed his words to Moses' mind, and the word was God. When Moses was asked if God spoke as a man, he said, "He is not flesh, but spirit, the light of eternal mind, and I know that his light is in every man."

Unfortunately, by the time most Jewish followers reading stories like this were no longer conscious enough to see the deeper meanings that earlier followers could. They became seen as more form-based, becoming more ego building than the ability to be ego dissolving, finding formlessness within the scriptures. Some stories still point beyond ego to some extent, others not as much. The Ten Commandments, for example, may have had much deeper meanings than more recent translations provide, but over time became transformed into the ego based god who is a separate entity to us. Several commandments view God as a very jealous individual, something that truly excites an egoic mind. It is difficult to know what is true and what is altered with any story from thousands of years ago. It is also impossible to see the world from the eyes of a human being that lived in that time. Within any egoic society, an egoic version of God can be the most useful for that time and place and certainly the most relatable to the unconscious mind.

When the idea of God is filtered through our imagination, the mind creates God in its own image. One example would be the need for God to take on the form of a man within a society that has become male-dominant through ego. "If men are superior to women, God must be a man," the unconscious collective mind thinks. Overlooking this unbalanced flaw that simply cannot be seen if the belief will not let it be seen. Moses apparently saw beyond this flaw as he connected with higher power, but this detail still becomes easily lost within a society that is not ready to look deep enough within. Leaving us with Moses using the word Man instead of Human within the current scriptures.

Looking at the burning bush story through the lens of modern physics. The idea of God appearing as a bush that is literally on fire but doesn't burn the materials of the bush. Creates an extra layer of skepticism toward its authenticity. Yet, in many ancient stories, some words are not meant to be taken literally. Could there be a simpler explanation? Some have suggested that the burning bush is a reference to the use of a psychedelic herb of some kind. Many ancient cultures have used psychedelics for many different purposes, usually for the sake of inner wisdom and guidance. When used wisely, they have been known to provide some amount of deeper understanding. It has only been in recent centuries that more and more cultures abolished their use out of ignorance of their potential benefits.

Whether or not any mind-altering experience was had, the guidance that Moses is said to have found could have only come from finding some level of inner serenity within the moment. Allowing him to not be fully taken over by what his mind was interpreting, and allowing new information to come into his thoughts. Is it possible that the bush was burning as a description of an awareness of Being, or living energy, noticed within a living bush during a meditative state of some sort?

Is it possible that Moses simply became still enough within to be able to feel as one with the plants surrounding him? Allowed Moses to experience long enough gaps between thoughts. Feeling the oneness of Being that flows through all life within himself and the plantlife for long enough to allow a deeper level of wisdom to come into his mind. Allowing him to see more clearly what to do in his current situation at the time. This would mean that the level of wisdom he received would depend on the level of stillness he was able to achieve at the time. If his mind was too restless, the guidance would not be as clear, and his current thought structures could get in the way of a deeper reflection.

In his first encounter with this phenomenon, he told God that clearly, he could not be the one to free the Israelites, that he was just one man. The experience he was having that would be described as talking to God told him that he could do it, that he would know what to do and what to say when those moments would come.

Whatever sort of experience this was with a burning bush, it seemed to have helped Moses to be more present with the challenges he faced within his journey. However, like any other human, we must be able to find enough depth or enough stillness within ourselves for intuitive guidance to come into this moment. These stories from the past may hold some degree of truth. How much is very difficult to say. They may only be stories. Either way, a story that holds depth within it can help point towards depth within ourselves, however truthful or fictional it may be.

Both the historical Jesus Christ and Siddhartha Gautama (the historical Buddha) are said to have gone into the wilderness to fast for about the same amount of time in order to gain the wisdom they were searching for. Jesus went into the desert to confront the Devil and was said to have fasted for about 40 days. The Buddha went into the forest and confronted Mara while meditating and fasting under a Rajayatana tree, also known as a Bodhi tree, for about 40 days. In both stories, they were tempted by the Devil/Mara until the temptations dissolved. These stories are most likely as similar to each other as they are because this has been an extremely common theme for our species.

Many humans have awakened beyond the ego stage throughout history. Most are likely unknown, while some have tried to point the way for others, each with various degrees of success. Many more have likely found it pointless, unable to put into their own words what will not make sense to anyone who simply hasn't experienced or become ready enough to experience it for themselves. Spending their remaining days either as a hermit in an isolated bliss or living amongst other humans who most likely simply couldn't understand their eternal bliss. Yet had to have benefited from it in various ways. Buddha is a title that simply means the awakened one. Gautama Buddha was said to have pondered this dilemma of either retreating or attempting to teach when he saw beyond the unconscious mind or Mara's temptation. He thought to himself, "No one is going to believe this." He felt he might as well just live as a hermit in the beautiful wilderness for the rest of his days.

He had been on such a long journey, both internally and externally. Trying any and every technique he could find within many schools of Hinduism in order to awaken as he journeyed around India. When he finally decided to simply sit under a Bodi tree, and let go of the seeking. When he finally found the simplicity of Being, of simply becoming still within. He thought no one in India, within the current spiritual communities, would ever understand how simple Being actually is. Yet it is said that the compassion he had found within his awakening would not allow him to go without trying. So he walked back into society to teach his unique expression of how to become awakened to those who might be ready. Buddhism has been one of the most efficient religions in preserving the original message of its teachings. It is still argued whether or not it even is a religion or simply a spiritual practice.

Buddha's method is the Middle Way, which is the opposite of the extremes he had found in many Hindu practices, the way most people were practicing them at the time. In ways that could not help but hold a mindset that keeps awakening as a future event instead of only being here now. The Middle Way points towards inner emptiness that transcends existence and non-existence, bringing the universe into balance within ourselves. Not being too hard or too soft, not being too much in form or too formless, and always looking for the path of least resistance. Finding the middle ground of any situation we are in, through always being in the present moment. His teachings are simple and profound for those who understand what he was pointing towards. His teachings were so effective in India that Hinduism would later become more realized by some as an earlier expression of what he was teaching. Yet, with the use of so many mentally perceived Gods, Hinduism can be cloudy in certain areas where Buddhism practices seem to cut through the darkness using our own inner light.

While the Buddha has been seen as a God by some, this idea remains a minority within the Buddhism community. This may be one of the reasons the teachings have been so well preserved. When the concept of God becomes involved in spiritual teaching, the ego's need to be good enough, as well as its fear of death, become the most important factor. Nothing wrong with this, but it fulfills ego needs with little to no ability to point beyond them. When ego makes worshiping a God the main focus, the mind's imagination creates cloudiness through misunderstanding teachings that cannot be grasped conceptually. The only way to truly experience Higher Power, or Collective Consciousness, is within. This has occurred again and again within most religions throughout human history. Continuously swinging back and forth between heavy ego interpretation to deeper spiritual understandings and back again. Somehow, Buddhism has, for the most part, managed to hold together the message that Sidhartha was an average human being who found within himself what any of us can potentially find within. On our deepest level, we are the eternal bliss within the now.

Jesus was in a similar situation, but depending on the gospel only had somewhere between one to three and a half years to teach before the religious/political climate of the region had him crucified. The Buddha awakened within a culture of Hinduism. Where the average person had some understanding of what "Enlightenment" was supposed to mean. Siddhartha was said to live to the age of 80. However, in Jesus' case, apart from the deeper wisdom that he was pointing towards within the Torah, for those who were able to see the depth. Jesus was basically starting from scratch within his community. Pointing beyond ego can be very scary for people still fully lost in identification with form. To the ego, not clinging to a set of belief structures in the exact ways they were previously taught can be extremely frightening. The same as death itself, the ego's biggest fear.

Many who had no idea what Jesus was pointing towards with terms like "Revelation" thought he was surely speaking blasphemy. As well as describing the end of the world, which, in a way he was. Except that when Jesus mentioned the end of time, or going beyond the world, something the ego would rather resist than try to understand more deeply. He was pointing to a world without attachment to thoughts or physical forms. A world with true freedom from suffering. Jesus was himself such a strong example of a human within the awakened surrendered state, that even his crucifixion can be seen as an example of a painful experience that lacked inner suffering. An example of how life is not without painful experiences, but suffering becomes optional once we know ourselves deeply enough.

Each Gospel tells the story of Jesus' crucifixion a little differently. It is incredibly difficult to say just how much of the Gospel's story structures have been altered over the centuries or how much is simply mythology, to begin with. Yet any mythological story can have little trinkets of helpful aspects when taken with a grain of salt, seen as allegorical more than fact. The story of resurrection is one that has repeated itself through many ancient cultures. In fact, many ancient Gods were claimed to be crucified, and most of them also claimed to have been born on December 25th. Usually having more to do with the winter solstice, when daytime is beginning to get longer again, than the actual date of birth for any of them. Historical Jesus is said to have more likely been born in the spring, but the winter solstice is a better date if you are trying to recruit people from more Earth-based religions into the Christian faith. Some ancient religions even depict their earlier God-like figure as the son of God who has been sacrificed before he is resurrected. Giving a believed proof of the supernatural kind that he is surely a God. Mythologies seem to reinvent themselves in one way or another as the cultural ego moves on.

This is also why early Christians chose early spring as the time of year that Jesus was crucified and resurrected. What better way to indoctrinate peoples of competing religions than to place new holidays around the same time as pre-existing ones. Resurrection merges well with the spring celebrations, when all the plant life is blooming again or being resurrected as the winter is coming to an end. Any story from our past becomes very fuzzy the more humans are motivated to manipulate other humans. The ego loves to create misinformation if it looks like it will serve the individual's desires. This being said, many people were crucified during the time that historical Jesus was alive. There is a strong possibility that he could have been as well, seeing as how unpopular he was with the status quo.

As we begin to see how ego patterns operate within ourselves. This crucifixion story within the New

Testament, at least before the resurrection aspects, can help point out the simple nature of the ego surrounding the crucifixion. It is up to the eyes of the beholder whether or not they find meaning or truth within the resurrection part of this story. There could be some truth in Jesus' followers feeling his presence in some formless way. Here, I would like to simply point out the historical Jesus' message of serenity during the crucifixion. He told his followers to forgive, not to seek revenge, but to forgive for the sake of their own well-being, as well as their deeper understanding of themselves.

For many Christians, the crucifixion story is more commonly known as an unconscious means to fuel their ego than it is known in a way that points beyond the imagined self. If the reader is not yet ready to recognize their own inner resistance patterns, they can more easily relate to the idea that they owe something to God because of this event. Giving the ego what it is looking for, the feeling of not being good enough. Negative ego reactions were present within most people who interacted with Jesus throughout his crucifixion. This can make it extremely challenging to notice how Jesus' lesson for his followers was him simply surrendering to what is as he was led back into nature on a beautiful day while also experiencing a painful death, which he had come to full acceptance towards. Knowing from a deeper place than his mind, this moment simply is as it is in a situation that he was not likely able to change the outcome of.

Maybe this is why, in the Gospel of Luke, Jesus tells the person on the cross next to him," Truly, I say to you, today you will be with me in paradise." Maybe he was referencing the afterlife, Heaven. Maybe he couldn't help but continue telling others how beautiful it is to simply be alive within this one moment. When we can fully accept the present as it is, a painful experience, as well as a moment of death (ours or others), we can still have a joyful feeling in the background. The joy of life is always within us, whether covered up by thoughts and emotions or not, as we are this joy, whether we know ourselves deeply enough or not. Jesus said to many people throughout his teaching, "Heaven is within you and all around you, yet you do not see it."

Jesus often spoke in parables. When he was asked why, he answered with a parable. Essentially stating that those who are ready to understand will see what the parable is pointing towards, but those who are not ready to see beyond the mind will miss the point entirely. This is the same method that is used in many other spiritual teachings. For example, Sufism, Taoism, and many forms of Buddhism. Words are used in such a way that they poetically point beyond what the mind understands, towards a deeper reality within us. In Jesus' Parable of The Sower, in the Book of Mathew, he says; "A farmer went out to sow his seed. As he was scattering the seed, some fell along the path, and the birds came and ate it up. Some fell on rocky places, where it did not have much soil. It sprang up quickly because it did not have much soil. But when the sun came up, the plants were scorched, and they withered because they had no root. Other seed fell among thorns, which grew up and choked the plant. Still other seed fell on good soil, where it produced a crop - a hundred, sixty or thirty times what was sown, Whoever has ears, let them hear." These words can be a metaphor for almost anything the mind can think of if the person doesn't realize what is being pointed towards. Yet, coming from a spiritual teacher who is pointing beyond ego, they can only mean one thing. Our thoughts take us in millions of directions until our attention is rooted deeply enough within. Only then can our spiritual awareness grow. Only then do we have ears to hear what is being said.

Identifying with a narrative that claims Jesus' death as being right while also paradoxically wrong in some way or another. Very easily creates an unconscious need to cling to the same negative emotional ego energy of those who needed Jesus to be crucified in the first place. The suffering that he was always pointing beyond. The same negative energy has made followers of many different religions throughout history certain that killing in the name of God was important for being favored by a mentally imagined version of God. In many cases, this idea has been used as an excuse for murder, pillage, and plunder as long as it's projected onto the defined "Non-Believers," those seen as unworthy of their mentally imagined version of God. In their minds, they will clearly be praised for the things they have done when they get to their imagined version of the afterlife. Yet never conscious enough to realize that heaven is already within them. The mind is simply thriving in the madness that is pushing heaven away. We could all be living in paradise, if we can just become conscious enough to realize we already are. All of Earth can be the Holy Land, and nothing would need to be fought for and easily shared. Once humanity finds heaven within ourselves, this is the next logical conclusion.

The reason Jesus said to "forgive them, for they know not what they do" as he lived his final moments on the cross. Is simply because he understood how holding a grudge is madness, creating problems out of simple situations for the sake of strengthening egoic madness. A mental identity is often most easily defined when the story of "me" is claiming victimhood. One of Jesus' lessons is to turn the other cheek. When someone slaps you on one side of the face, turn your head to allow them to slap the other cheek if they choose to. This does not necessarily mean to let people walk all over us. It is likely that it wasn't even meant to be taken literally. Instead, it was likely meant to simply display the importance of not taking life too seriously, regardless of the situations we encounter. Allowing serenity to become more important than the madness that is created when holding onto a grudge or resisting what already is within this moment.

Like most religions, Christianity can be a tool for fueling various degrees of egoic suffering. Yet can also be used as a tool for awakening beyond suffering, depending on our level of consciousness. Many people can have a hard time getting back into a religious/spiritual belief structure that they only knew an ego version of at an earlier time. There is no one path that is right for everyone. Yet, at the least, simply noticing deeper layers to stories like these can be more than enough to help us understand how we are all moving along the same collective journey of humanity's evolution. The same evolving consciousness expresses itself through individual human forms. Possibly helping us to understand more deeply that we are the only person who can walk our own unique path of awakening. While others might guide us, only we can liberate ourselves.

Today, as we journey the wilderness, we are no longer a species that lives in separate isolated areas of the world as much as our ancestors. Only able to access the stories of one to maybe a few cultural narratives. We also are no longer at the point in our collective evolution where finding the inner light is reduced to a few random individuals. A new spiritual leader is no longer necessary as we've reached a critical threshold of our collective ego. We have a stronger threshold of knowledge about other cultures that our ancestors had no access to, and we are more ready to awaken collectively than our ancestors were at any earlier time. Extreme fasting, or depriving ourselves of surface pleasures to any extreme level, is not as necessary as it might or might not have been for certain individuals in earlier times.

These extremes are probably more likely to cause problems as opposed to the simple practice of noticing how the mind clings to certain habits. You may find that giving up certain habits is necessary, as well as starting new ones, like meditation, which is very helpful. If this book is written well enough, I can only point towards what awakening is. It is up to each individual to become aware of what is needed for our own unique journeys. You have all the readiness needed to awaken at this moment.

Striving to find peace within through extreme practices like starving ourselves and staying out in nature for a month or two without food will most likely feed ego elements that still need to get to an imagined future place instead of simply being here now. Yet finding a balanced or middle way that works for us allows presence practice to remain joyful as we become more aware of the simple depth that has always been within us. Taking a hike from time to time out in nature can be a simple way, a balanced way of practicing presence, while other forms of life help us feel the one consciousness within. Being out in the natural world can be very grounding in many ways. One of which is being surrounded by life forms that lack an unconscious human mind can help root us in Being. As thoughts continue to come in at these times, these experiences can help us see how the ego is clever and can be tricky to move beyond. Yet presence is extremely simple, as simple as nature. If the mind is lost within negative complexities, the simplicity of knowing yourself more deeply begins to give the ego's game away more easily as the present moment moves on.

Today, as we walk into the wilderness. Simply hiking on a trail or sitting on a park bench can be a very simple practice in finding our true nature, our deeper reality. Anything we give our full attention to will have more depth than when we are lost in thought. Anytime we give this moment our full attention, the present moment is infinite. There is no time or mental scale of how deep we may or may not be, how deep we still need to go within self awareness. This moment simply is as it is. Presence has no limited depth. Only the thoughts of awareness can be limited. A limitation that cannot last if your awareness has come far enough to understand what this book points towards. However aware we have become is irrelevant. You are the present moment, experiencing itself as a human. Every time we fully accept the now, this truth reveals itself and becomes experienced a little more deeply.

It is important to realize whatever depth we experience within nature, is not something we can only find in nature. It is something that is not a "thing" but an inner spaciousness that has always been within us. Suppose the mind is able to convince itself that only the wilderness can produce that level of inner depth. This can fuel thoughts that create a problem out of needing to be there instead of simply being here. Not getting enough of a conceptualized version of nature. If you notice this mental pattern, you haven't failed. You've succeeded in noticing a pattern of pointless drama within the mind. What is made present can no longer operate on an unconscious level. Going out into nature can be very helpful, but it is not the main ingredient for awakening, simple inner awareness is.

While nature may be limited in the cities and suburbs, there is always some nature to be found wherever we are. Houseplants, gardens, and trees along the side of the roads. City parks can be a great place to go on a regular basis to practice presence, notice our breathing, and feel the inner body. Even in the middle of a city filled with skyscrapers, there is still the sky. Simply taking a moment from time to time to be still with whatever surroundings we find ourselves in brings us back to the nature that is within us and reminds us that this human form is nature. Other natural biological forms are helpful, but becoming more aware of our natural form is the most important and the main source for becoming more aware of our inner light.

One could think that sitting in an airplane would be the farthest away from nature we can get. Especially how stressful it can be for the mind to go through the process of getting to the airport, through security, waiting at the gate, and filing onto an airplane. Along with many other stressful human minds that have thought about how stressful this experience will be for many stressful hours, days, weeks, or even months leading up to this overly thought-about event. Simply walking onto an airplane has been compared by many minds to being crammed into a can of sardines. As the large group of people tend to move slowly to their seats, an impatient mind finds more and more things to complain about.

Many people lost in repetitive thinking will completely overlook the positive aspects of this experience. So lost within the mind's need to get to a future destination. They can easily overlook how quickly an airplane is getting them there. Much faster than any of our ancestors could ever imagined traveling. Stress from not wanting to be here will easily turn into complaint after complaint about how long the trip is taking. So lost within a mindset of resistance, that they are unable to notice how amazing it is to be traveling at 30,000 - 40,000 feet above sea level. If you've ever sat in a window seat on a plane, you probably had an indescribable experience as you looked at the Earth from that position in the sky.

When I was young, I got to experience the airline industry from a unique perspective. My mom worked as a gate agent for over 20 years, allowing our family to fly Standby. Which in many cases means you stand at the window and wave goodbye to a full aircraft. Yet when there were empty seats, we could fly for free anywhere within the continental United States. This allowed us to take trips to places like Washington, D.C. several different beaches in Florida. As well as visit family in California, Colorado, and Minnesota.

We also had the privilege of encountering the stress most humans take with them to the airport in their need to be "there" instead of simply being able to be at the present step needed to get them "there." Flying standby can add an extra layer of stress as you only get on the plane part of the time. That part of flying was definitely stressful some of the time as a kid, not yet realizing that all my stress was coming from my thoughts about the situation, as opposed to the situation itself. This could have become a strong practice of presence if I understood at the time. Yet, reflecting back, all of life's seemingly stressful situations were always simply teaching me more about myself. The egoic mind can't help but create stress for itself, always looking for the next thing it thinks it needs to stress about. Because of this, these patterns are always giving themselves away to some degree. If we are not yet ready, stress is preparing us to become more ready.

However stressful the airport might have been, and my mom has many stories about working in that environment of stressful minds. As soon as we started taking off into the sky, it was a totally different experience. On the one hand, you're simply sitting in a seat within a cylinder-shaped human-made structure with wings. On the other hand, if you look out the window, heaven is literally being displayed from about 30 - 40 thousand feet above sea level. There are no real words to describe this, and luckily, as a kid, there was still enough wonder in life to be in awe of the indescribable. While not needing a mental label to be put in front of it. These may have been my first glimpses beyond the mind without fully realizing it.

All of us have these moments in some way or another growing up. When our culture is informing us to be more conditioned, while something deeper inside, not fully realized yet, is pointing us beyond the madness of human conditioning. It is the nature within us, the timeless essence that is in all beings. There is nowhere this cannot be found, even if it's covered up by overthinking. There is no thought without the deeper awareness that allows thought to exist. The more we realize our own depth within, the more going into nature becomes going into a deeper self.

When something is lost for a long time and then found again. The newly found appreciation gives it a depth that was not there before. Consciousness wants to become more conscious on this planet. It requires the process of becoming so absorbed by the mind that when we come back to what consciousness really is, our true nature, it is experienced with a depth that could not have been truly found until it appears to be truly lost. So, none of us are truly lost along this path. We can only appear to be.

Some roads whined more than others, but eventually, all roads will lead to Zen. Not by all who are alive today, but at some point, it is inevitable that all of humanity awakens out of ego. When our consciousness is no longer lost in our mind, we experience the profound simplicities of nature more clearly and directly. We can understand from a much deeper level what the 13th-century Islamic poet Rumi meant when he said,

"You are not a drop in the ocean. You are the ocean within a drop".

EVOLVING EXERCISE

Going out into the wilderness is a very helpful exercise for practicing presence. What can be helpful on a more regular basis, is to simply consciously acknowledge the nature that surrounds us within our daily routine. Simply taking a moment from time to time in order to be still with a houseplant, for example. Practice being with the plant more fully, as opposed to only thinking while near a plant. Practice being more aware of the plant life surrounding the city streets, as opposed to not even seeing them at all when we are fully engaged in thinking. Maybe spending a short time in a park, during a lunch break, for one example. As well as working in a garden from time to time. Whatever we give our full attention to becomes a tool for moving beyond the overactive mind.

Many times, something just as simple as looking up into the sky can be extremely therapeutic for letting go of overthinking. Especially around sunrise and sunset. Anytime throughout our normal daily routine, when we simply pause for a moment and become still with whatever amount of natural surroundings we find. The simplicity of nature can help us realize all the truly meaningful things in life, like joy, love, beauty, and inner peace. All can arise outside of the realm of thought, within our deeper essence. At any time that we simply become still enough within, our natural state of Being brings us back to the truth. The simple yet profound bliss that can always be found to some degree within this moment.

Turning a Problem Back Into a Situation

"Yield and overcome, bend and be straight. Empty and be full, wear out and be new. Have little and gain; have much and be confused."

-Lao Tzu; The Tao Te Ching

The ego will very willingly cling to a negative identity so long as there is a mental identity for it to cling to. Within the Ape Kingdom, there is some evidence of earlier or more primitive forms of ego. The alpha male within a Chimpanzee family, for example, will wake up in a bad mood from time to time since no one can disapprove of or disagree with the alpha male as he is the most physically fit and dominant. It is always best to simply stay out of his way when he wakes up in one of his moods. He can be physically abusive within this moment if someone happens to be in his way. This dense, negative mental/emotional state is one of the best ways to strengthen an ego identity. It is essentially what ego is, a certain heaviness of self. The only difference is that the Chimps will get over it quickly. Without a need to cling to the mind's repetitive narrative about or around the situation, there is no prolonged inner heaviness afterward. When we observe Chimps in their natural environment, it becomes clear that they can be extremely peaceful most of the time, with only short bursts of violent behavior of any degree. As far as we know, a Chimpanzee will not identify with a bad mood for any longer than the mood actively exists.

Another example of aggression, where it seems to hold a more utilitarian purpose, would be when several members of one group of Chimps are battling another group for the sake of territory and resources, like where fruits are growing. Within these battles, their energy becomes very heavy, dense, and aggressive. Yet once the battle ends, they become peaceful again. The patterns of these moods obviously repeat themselves from time to time, creating some level of familiarity. Holding the role of alpha male, as well as maintaining territory for the group in a military sense, must hold some amount of conflict within the individuals while trying to keep control over these elements in an ego-like way. Meaning it is likely they are maintaining a mental and emotional identity to a certain extent.

Yet Chimpanzees appear to be just below the level of overthinking which creates the full aspects of ego, fully identified with thought. It can seem like their level of consciousness is partially in ego and partially pre-level of egoic consciousness, as their minds sway in and out of these egoic states. If this is true, this is the opposite of where an increasing number of humans are finding themselves today, partially in ego and partially beyond. The difference is that any human finding themselves within this transition doesn't have to wait any amount of time to continue evolving. We have everything needed right now to become fully awakened beyond the level of ego. Chimps may need countless numbers of generations in order for their species to physically build the mental capabilities to become fully within the ego stage of consciousness. Humans have obtained all of the mental aspects. Now, our challenge is all about formless awareness within ourselves.

For Chimpanzees who are likely slowly physically evolving into identification with the mind, the present moment is still more relevant than any story of time that the mind may hold. Throwing a negative tantrum from time to time might help a Chimp in various physical ways; perhaps the alpha male benefits most from being the most egoic. Yet this state cannot take them over to the extent that coming back to the calming balance of the present moment is fully forgotten. Once our earliest Himinin ancestors reached the level of thinking, that brought in a need to fully identify with the mind's stories of time and the moods that can accompany it. Throwing a negative mental/emotional tantrum became a very helpful way of strengthening ego regardless of any social role or life situation one might find oneself in. Creating the next challenge for the growth of consciousness on planet Earth. The ability to have a complex mind and not be fully lost within it.

It is unknown if the Tao Te Ching (The Way of Virtue) was written by a single individual or if the original texts that date around 400 BCE had accumulated from a number of people over time. The name Loa Tzu simply translates to "Old Man" or "Old Master" in Classic Chinese. Taoist tradition tells the story of an elderly sage who became fed up with the insanity of his fellow humans and decided to spend his retired years in solitude. So he climbed onto the back of a water buffalo and rode westward towards what is now Tibet. Along the way, a gatekeeper tried to persuade him to turn back or at least create a written record of his teachings. After agreeing, Lao Tzu came back just a few days later. Presenting him with the original 5,000 characters of the Tao Te Ching.

To yield and overcome is a presence practicing method Taoism has taught for many centuries. To simply notice when our doing/thinking becomes dysfunctional. Then pausing, or becoming still momentarily, in order to find a more balanced way to continue with the current situation. Yet, this wisdom is easily overlooked when continuous thinking is all we know ourselves to be. To stop following an established train of thought is something an overactive mind rarely wants to do. Unless it can think about this as an idea for a long enough time to keep thoughts going. However, simply noticing the mind's inability to stop thinking is to yield from a deeper state of awareness, allowing a deeper place within to become recognized. The gap between thoughts might be very short as we begin this practice.

Yet any noticeable space is a profound shift from no noticeable space.

It can be challenging enough to yield from thought patterns that are determined to think. When the emotional ego becomes involved, it can seem impossible. That is, it can seem impossible according to the next thought and the next thought. In certain ways, the emotional ego is like the mental ego's shadow. You could also say that the unconscious mind and emotions are like two stars that orbit around each other. In many ways, the gravitational patterns of one cannot exist as it does without the gravity of the other. Our overall ego experience needs both mind and emotion in order to hold the illusory sense of self together.

The one aspect of Taoism that most people know of is the Yin Yang symbol. The circular symbol with two polarities, usually represented by the colors black and white, with a smaller circle of the opposite polarity on each side. Demonstrating there will always be a little of one within the other. Light and dark, feminine and masculine, conservative and liberal, and so on. All opposites tend to have more in common than they may first appear. Which color represents what side seems to change depending on who you ask. Making the symbol an even better representation of the paradoxical nature within our universe.

Yang can be seen as white, representing form or action. Making the Yin black, representing the formless, non action, or Being. To be too far in either direction is to be unbalanced. Yin and Yang can also represent the relationship between two opposite expressions of form. Male and female, cold and hot, beginning and end, happy and sad. Thought and emotion may not exactly be opposite in certain ways. Yet can be seen this way as we notice their interactions with each other. Their sort of marriage with each other.

When the mind resists what is, whether in the past, present, or future, there is always some level of negative emotions involved. The opposite is also true. Whenever a situation is embraced in a positive way by the mind, positive emotions are always involved in some way. On the other side of the spectrum, certain emotions will often spark repetitive, conditioned thoughts that complement the emotional frequency. Going back to the chicken and egg analogy, it can be extremely difficult at times to know which comes first. As they are so well entangled, they can appear to occur simultaneously. More important than determining which expression sparks the other is simply becoming more aware of the full spectrum.

Complexities of thoughts can, of course, become lost in conceptual explanations. Making it tricky to notice exactly when our thoughts have become dysfunctional if they can be noticed at all. Becoming more conscious of the feelings within the body can simplify everything when we bring a higher awareness into emotions that are resistance-based. There is a potential to recognize them more deeply for what they are. Emotional frequency not who we are, not the situation we are encountering, but simply a physical experience within the body. We begin to more easily notice these patterns for what their purpose is. Like other senses in the body, emotions are simply a way of recognizing the world around us and the situations we find ourselves in.

They are a way of understanding and relating to other people, as well as other life forms. Emotions that are not fully conscious, that we identify with as who we are become very useful for the survival of the ego. The survival of dysfunctional feelings about situations, ourselves, and others. As well as feeding and being fueled by dysfunctional thoughts about situations, ourselves, and others. The mind can argue the importance of a means to an end or its opinion about a situation with a certain blindness towards any dysfunctional behavior; however, if we can more consciously notice when thoughts about a situation project certain feelings. Simply noticing an emotional response can more accurately determine whether our thoughts are useful or dysfunctional. These emotional reactive patterns may not be noticed right away as we begin practicing presence. Yet the more we notice, the sooner we become able to yield and overcome any ego reactivity.

On the other side of this phenomenon, when repetitive emotional patterns seem to come from out of nowhere. The repetitive thought patterns they tend to feed can be what gives away the habitual conditioning of both, as the ego needs identification with both thoughts and emotions in order to sustain the illusory self-image. Becoming more aware of one helps us become more aware of the other. Becoming the watcher means that there is nothing that needs to be done. What you seek is within you. Changing some patterns of how you think and approach situations can be very helpful. However, these patterns will balance themselves through your awareness of them. When the light comes into the darkness, there can be no more darkness.

This is why humans seem to always create the same problems for ourselves over and over again. An important level of inner depth is required in order to find balance. Identifying with inner resistance is what keeps the emotional ego alive. There is no positive without a negative, only when we transcend this particular yin-yang by becoming aware of our deeper, formless essence, can we experience both without being lost in clinging to one, while resisting the other. Sometimes, we cling to positivity while resisting negativity. Sometimes we are lost in negativity, while resisting the positive through disbelief that we could find the positive. A need for things to be positive can be a form of resistance simply because resisting resistance is resistance.

To be accepting of whatever emotions we feel at this moment. Is to be free from the need to be lost within any particular feelings, allowing us to experience emotions more fully. Allowing us to experience them without the mind getting in the way of an interpretation of how they should or shouldn't feel. As we become more aware of the depth within us, thoughts and emotions are seen increasingly as a surface phenomenon. No longer pulling us in every direction that is suggested. Seen more as the tools they are meant to be for interacting with the world around us and less as who we are.

Just as every mental ego is its own unique expression of habitual patterns, the emotional ego is a unique expression for everyone. Our conditioned thought patterns, our family, friends, cultural environment, as well as genetics are what bring about how we experience identification with positive and negative emotions. As mentioned before in an earlier chapter's reflections on warfare, traumatic events can create emotional reaction patterns that basically consume the mind's normal thought patterns in various ways. Creating a lower level of consciousness and a higher level of suffering until these specific emotional patterns reside. Usually having active phases and dormant phases that repeat themselves to various degrees, with unique timeframes for any person.

Experiencing war between nations, as well as other forms of extreme physical trauma, is one of the ways to acquire an identification with dense negative emotional patterns like these. Yet, there are many ways that an emotional ego can grow to this level of intensity if our home environment while growing up is filled with drama, like having parents who are always arguing and fighting, for example. If our neighborhood/city environment is heavily impoverished and therefore filled with many types of drama. These patterns are also embedded within our DNA, as they have built up for as long as they have within our species. Making an extreme level of unconscious emotional patterns easier to acquire than they may seem. Simply going through puberty can trigger ego patterns that are simply waiting dormant within our genes. Waiting for the heightened level of hormones to trigger them into action. One of the reasons depression can be so strong within teenagers.

When negative emotional reactive patterns become intense enough, the resistance towards negativity, as well as pain, can become flipped into being unconsciously seen as a positive or even viewed as pleasurable in a certain way while these patterns are active. This flip occurs because the mind's identity is essentially taken over or consumed by these negative emotional patterns. This creates a mental addiction to the negative emotional pain, as a part of the mental identity becomes invested within the experience. The repetitive patterns of pain become seen as a part of who we are. At the same time, the emotions themselves become addicted to the negative thoughts that continuously fuel them. If these patterns are intense enough and or last for long enough, both thoughts and emotions create an unconscious need from each other for this level of intense negativity to continue. Making the pain seem pleasurable in a certain way by both the mind and emotional body whenever these patterns are active. This can create a need to inflict emotional or even physical pain onto ourselves, as well as others, while in this extremely unconscious state.

The amount of time an active cycle of this intensity occurs depends on the person. Some patterns may last a few minutes with extreme intensity. Others may last for hours, days, or even longer with less intensity most of time. An argument with another person could be a good example of short-lived bursts of strong intensity with this unconscious addiction, with a longer-lasting, less intense afterglow as the mind replays the argument over and over again. Within an argument, some people may slam their fists on the table or throw an object furiously during these short-lived fits. Another example of emotional pain addiction would be holding onto a sort of self-degrading inner narrative that is charged with negative energy. Often, with a lower intensity of resistance, that can be active for much longer than a short lived reactive burst can be. Both of these examples can go hand and hand as no dormant cycle is ever fully dormant within someone who is fully lost in ego.

At whatever level of resistance the ego may currently hold, complaining about situations or other people can play an important role. Continuously reestablishing the importance of resisting what is here now for the sake of inflating a sense of self-image through playing the victim. The only reason we complain is to reestablish how seemingly victimized the mind declares itself to be. Not to be confused with real issues of physical victimization. The ego's victim character will play this role regardless of the external circumstances. All that is needed to move beyond these experiences of inner resistance is to simply practice noticing them when they occur with the least amount of judgement that we can have while noticing.

Many people experience both long-drawn-out self-loathing to some degree as well as short bursts of mental/emotional chaos to some degree. When the emotional patterns subside, the mind's attraction toward the negative emotions subsides as well. Yet, without enough awareness of self, the identity remains, planting the seeds for the next active cycle to begin at some point. Usually, this creates a mental resistance towards these unconscious patterns while in the dormant phase. Which unconsciously creates more negativity for when they become active again. What we resist will always persist.

Dense emotional ego patterns like these require an active phase and a dormant phase in order to survive as well as thrive. They become active again anytime they are triggered by something, as well as when they have not been fed for a long enough time by mental negativity. In the dormant phase, these patterns are never fully dormant but interact with thoughts on a smaller scale. Being fed by just enough negative thought patterns to stay as a part of the identity. Then, when enough time has passed, and or just the right mental/emotional trigger occurs, seemingly out of nowhere, a new active phase begins. Consuming the mind with extremely intense patterns once again.

Like all ego patterns, they can only repeat these cycles until they are made conscious enough. Also, like any egoic pattern, once we recognize them for what they are, they cease to be ego. They are no longer as unconscious as they were. The patterns may still repeat themselves for some time, depending on the momentum they have. Yet the simplicity of presence allows these emotional patterns to balance themselves out. The mind has a harder time re-identifying with and renewing patterns we no longer confuse with who we are. Like the mental ego is time-based, the emotional ego is completely past-based. The more we introduce presence, the less the illusion of time can take us over. The patterns of the emotional ego will likely be painful to face fully at first, but this, too, is a part of the past pain reestablishing itself. The more present we become, the more these patterns are transcended into presence, into peace. All that is needed is time to become more present, transforming inner resistance into inner peace. The historical aspects of our lives still exist and hold a certain importance. We simply no longer need to cling to a story of time that is no longer here now in order to define ourselves.

In my case, the emotional ego thrived simply because the mind's thoughts about "my life" became negative enough to allow a strongly negative relationship between the two to find a seemingly endless cycle of suffering. Nothing in my childhood was ever extremely dramatic. It wasn't until depression came along around the age of 13 or 14, through the body's natural increase of hormones, that the emotional ego began to truly strengthen itself. The mind and the emotional body suddenly became increasingly attracted to each other's production of negativity. I wasn't negative all the time, but enough to create an identity around negative thinking. Not that I would have been able to explain it this way. Depression seemed to come in several waves of this unconscious attraction toward negative thinking. Every few years or so the emotional ego would gain its density, making the experience stronger. Then, stress headaches started occurring in my late teens, becoming stronger and more frequent through the mid-twenties. Influencing more stress, as well as a need to always resist the thoughts of the next headache, or simply resisting the next situation that might be perceived as stressful.

Thoughts about myself were the biggest culprit of my mental dysphoria. The mental "I" became amplified as the mind needed to tell itself how bad I felt about myself without being able to see that there are two selves within that sort of thought: the mind and the observer (how I felt about myself). I obviously could only know myself as the thoughts themselves. I was often unconsciously attracted to looking to see how bad my life was, how estranged I felt within the mind's story of self, and how different I felt from other people. I always felt that other people must be doing things in their lives much better than I seemed. I also seemed to never get anything right, according to my thoughts. When I did feel like I might have done something right, it wouldn't be too long before the continuous negative thinking would find some reason why it also wasn't right enough. Always feeling the feelings of being wrong and strengthening a self-loathing expression within the experience of ego. The conceptual identity could never truly create for itself a confident persona but thrived as a persona of low self-esteem.

This established identity wasn't always intensely negative, but negativity was a recurring theme in various degrees. As a child, there were some small aspects of this, but nothing overwhelming. As a young adult, there were plenty of times when I found enjoyment in one way or another, as well as plenty of moments where the heaviness of the mental/emotional ego became active again, reinforcing its need to suffer. Like most people the mind's outlook would swing back and forth from good to bad. Yet, on a longer scale, the ego was slowly strengthening itself as time moved on, strengthening stress headaches as well. Negative thinking was strengthening the idea that "my life" was never quite in the right place. Without realizing it, the need to get away from myself was exactly where my mental identity wanted to be. Ironically, this may have been a part of the ego's downfall. By holding an identity of needing to get away from myself, the seed for looking deeper than ego may have been planted.

At the age of 26, I started falling out of bed early in the mornings from time to time and dislocating my shoulder due to the development of Grand Mal Seizures. At this point, the inner resistance had reached such a level of intensity that it helped the increasing stress headaches transform into seizures, with the likely help of certain genetic aspects. Both inner suffering and seizures together are likely what began forcing me to question more deeply what I was experiencing within the mind, just slightly beyond the thoughts about what I was experiencing. Without understanding it, a deeper level of awareness than the mind began to emerge within this life. However, it would be a couple of years before this would become more understood through reading Echart Tolle's work. Those two years were likely very important for the process of awakening. It probably took this length of time of the ego's suffering continuing to stumble over itself at such intensity for me to be ready to understand what Eckhart's teachings point towards and begin to truly shift beyond the suffering.

The first few times I was waking up after falling out of bed, doctors diagnosed me as having night traumas. As I went to the emergency room several times after dislocating my shoulder, they couldn't understand why it was being dislocated backward as it's much easier for the shoulder to come out of its socket in a forward motion. One day, I passed out at work and woke up in an ambulance, being told I just had a seizure. They realized I was falling out of bed while convulsing with my arms pressed tightly against the upper abdomen. When I would roll out of bed and hit the floor, the intensity of the convulsions meant the upper arm would be forced around the Acromion, the part of the shoulder that holds the arm in place. Luckily, I learned how to push the arm back into the socket by leaning against a wall at just the right angle. Epilepsy, however, has been something I still hope to find a cure for someday.

So far, I have found various degrees of success trying many different medications. Modern medicine can treat the symptoms, but so far, it has not learned enough about the complexities of the human mind in regard to seizures in order to be able to cure most people. Children can be the most successful, as their minds are still developing and haven't created these patterns for as long as an adult mind. However, science is always progressing, and it is very likely that a simple cure for epilepsy will be found at some point in the future. I am simply grateful to have found the medications that work for me as well as they do. I have luckily experienced a great reduction in strength when I do still experience a seizure. I feel a part of this has been the reduction of stress in my daily life.

Anyone who develops epilepsy has different causes, and genetics are usually a key factor. Yet, as I have become more familiar with how a seizure feels, just before blacking out, as well as afterward, it became noticeable that for me, negative mental and emotional energy was a large part of what helped create this disorder. If not, resistance patterns were certainly opportunistically feeding off the situation before I became aware enough of this. In my mid-twenties, without me realizing it, the many dysfunctional qualities of this new phenomenon of seizure activity. Had started reflecting and more easily displaying emotional ego patterns that were now more intense than before the seizures began.

I became metaphorically crammed into a corner with nowhere else to go but to look more deeply within. Then at seemingly the perfect time, I began finding spiritual teachings from Eckhat Tolle, as well as many other teachers of today as well as throughout history. Some of which I knew of, but was now seeing in a new light, like the Buddha, Jesus, and so on. As the Tao Te Ching says, "When the student is ready, the teacher appears. When the student is truly ready, the teacher disappears." The second part means that we only need the pointing of others for so long before we know our own path of awakening.

With a renewed and deeper understanding of myself, I began accepting/using the situations I encounter in life to move beyond the ego that fuels itself by resisting these situations. Allowing me to begin moving beyond what I would call the worst part of a seizure. Not the physical attributes, which can be pretty painful, but the inner resistance towards the situation. Now, being able to see myself more deeply, I began to practice yielding in order to overcome. I began to make use of an extremely dysfunctional situation for the purpose of transcending beyond the suffering that very likely helped create the situation in the first place.

Simply by accepting this one moment as it is, when a seizure occurs, it already is as it is. If the mind can't accept that a seizure has occurred, this too simply is as it is and will not last for more than a moment or two. The more I would simply watch these situations, the less I would identify with a passing event, and the more easily I could observe reactive patterns from a deeper place within. On one level, I am a person with epilepsy. Yet, on a deeper level, I am simply the present moment, experiencing itself as a human being who experiences seizures from time to time. This practice, of course, helps with any situation that the mind resists. Flipping what would strengthen ego into an exercise of dissolving it.

The human level is still an important part of who I am, just no longer seen as all that I am. Without a need to identify with the epilepsy label, life becomes as simple as what is within this one moment. Suddenly, both seizure events and mental/emotional reactions became my teachers for awakening beyond ego. This is how it has occurred for me, but the journey beyond ego is different, as well as possible for everyone who becomes ready to look deeper than the level of thoughts. By no longer clinging to what was or an imagined idea of what might be, we become free in any life situation we find ourselves within. How liberating it is to no longer need the present moment to be any particular way, fully embracing what is here now.

I have even come to a point where I am grateful for the fortune/misfortune of epilepsy. Suppose the mind had never come into such extreme aspects of suffering that seizures enhanced. I may not have been able to begin moving beyond the suffering. Perhaps the strength of egoic restlessness would have stayed at a level just under the threshold needed to begin looking more deeply within. It is hard to say what would be different, and there's no way to know. Either way, this is the path I've found myself on. The real miracle is when we are able to fully accept the path we are on, not without making choices but no longer making dysfunctional choices through a dysfunctional state of mind.

Not everyone needs this extreme threshold of suffering in order to shift beyond ego. We can more deeply realize ourselves in other ways as well. Suffering has simply been the most likely way so far. However, as people continue to shift collectively, it becomes less necessary for us to suffer in any extreme way as there will become more of a collective example of people living beyond ego. Today, there is more of a collective example of humans within the ego state, but when this flips, there's an increasing amount of awakening, as well as awakened humans. It will be easier for more people to learn about suffering with little to no need to be submerged within it.

167

Whatever it is in life that brings us to this point of looking beyond ego, that is truly a gift in many ways. When we can accept life situations as they are. They will always bring with them the gift of finding inner serenity, the main ingredient of inner peace, and acceptance towards what already is within this one moment. I would still love to be cured of seizure activity at some present moment in the future. Epilepsy still creates limitations and challenges in my life. Yet, I no longer suffer from seizures. If/when they do occur, I simply live with what is here now. Always turning a problem back into a situation.

"If you understood the nature of suffering, you would not have to suffer. You must first understand what causes suffering, and then you will have the ability to transcend it."

-Jesus Christ

The Hymn of Jesus from The Acts of John

Karma is a word that easily gets misunderstood due to superstitious belief structures. Its original meaning is simply "action". Often used to imply that every action has some sort of consequence, whether it's positive, negative, or neutral. One movement of energy will often lead to another. On the surface of our lives, this can be as simple as What happens if you pay your bills? What happens if you don't? What happens if you get to work on time? What happens if you don't? Karma is neither good nor bad, but simply one action/non-action that leads to results from that action/non-action.

On a deeper level, a mind that doesn't realize how much it is unconsciously attracted to negativity. Karma can not only be turned into sustainable suffering for the remaining life span of one person. A person who unconsciously creates suffering for themselves unconsciously imposes these lower frequencies of energy onto surrounding family, friends, co-workers, and any other human they come in contact with. It depends on the level of awareness of these other people for how much of this negative energy they might unconsciously absorb. Parents with mental and emotional suffering can't help but teach their children to view life within the same context of their suffering. Passing this karma onto the next generation. However, suffering within our species has been with us for such a long timeframe that we are all born with some degree of this karma within our DNA.

It can be easy to blame our parents for this karma of suffering. Yet this only allows the patterns to continue repeating themselves through this resentment until they are made fully conscious by us or a later generation. This karma remains until the time that someone becomes ready enough to fully forgive their parents, as well as the influences of other people's suffering. No one suffers on purpose, and no one spreads their suffering to others on purpose. The mind may think that their parents just didn't have the right moral values. Thinking, " When I raise my children, it will be different". Then, when they find themselves in the same parenting situations, they may or may not become as unconscious as their parents, but for unexplainable reasons, they can't help but act in very similar ways, even through choosing different strategies of discipline and overall interactions with our kids. We can only ever be as conscious as we are. The less conscious we are, the more dysfunctional our actions are.

It is not necessarily positive actions that cancel out negative karma within us. It is the deeper understanding of ourselves and what we are experiencing that allows truly powerful change. Positive action will likely be involved, but only when it is rooted in a deeper place than thoughts and emotions can it truly heal the wounds of our ancestors. When we are conscious enough of ourselves, positive actions no longer need to be micromanaged. They will simply come more naturally.

Much simpler than a mental analysis of what is right or wrong, good or bad, positive or negative. Is the deeper experiential understanding of light and dark. The mind might see the situation of millions of generations building up to this life's current expressions of suffering as pointlessly impossible to dissolve within one lifetime. Surely, no number of good deeds can make up for all the actions throughout history that have been rooted in madness. Except, this implies that what we need is time.

Conceptual time is what created this darkness. It is the lack of time which dissolves it. The truthful light of your presence within this one moment is all that is needed to dissolve the time-based darkness that pretends to be real, as there would be no thoughts or emotions without a deeper consciousness for them to exist within. It is simply through becoming aware of this deeper consciousness that exposes the darkness to the light. No unconscious darkness can exist once we shine the light of consciousness onto it.

As you become more aware of your own patterns of suffering. You will become more aware of the eternal peace that is always within. You may even come to a point where you can't help but be grateful for the darkness that has created the need to look more deeply within yourself as you become more aware that it is nobody's fault that we experience inner suffering but a natural process of an ever-evolving consciousness within the universe. Life has a way of giving us what we need in order for our consciousness to evolve.

Finding this message within a book can be extremely helpful. However, it is the experiential lessons that provide the most growth in a deeper understanding of ourselves. Planning a conceptual strategy of how not to react negatively within future situations can be helpful to some extent. When these patterns do arise, they will likely still take the mind over in a way that does not allow us to strategize through the mind that is lost in these habitual reactive patterns. Only in the present moment, when these unconscious patterns arise, can we begin to observe them firsthand and begin to move beyond them. Only in the present moment can we practice yielding to overcome what is ingrained within our mental and emotional karma.

Unconscious habitual patterns are like the weather. Unpredictable to some degree when they may or may not present themselves. Yet, they usually express the same familiar patterns when they do. Within our daily routine, stress patterns present themselves in many ways, some subtle, others not as much. A good example of physical expressions of stress can be touching our faces. When we find ourselves placing one, or sometimes both, hands on the forehead, for example, maybe even rubbing up against it. This tends to be an action that is an expression of stress. Notice the facial expression that comes along with this action.

A simple action that seemingly comes from out of nowhere, promoting the need to stay within the identification of a negative state of mind. This is not to say that every time we touch our face we are repeating stress patterns. Simply, any action that feels like it is charged with a certain heaviness or negativity is a pattern of ego-clinging to an impulse of resistance to what is. It's not the action itself but the energy behind the action that needs to be made present. Not being able to be physically still is another strong indicator of inner restlessness. Like nervously moving around in a chair for one example. The challenge is to not resist resisting and learn to simply watch how these unconscious stress patterns operate within us. There is nothing that needs to be done in order to shift beyond stress patterns. Simply watching them is the profound shift of consciousness already occurring within you. To see the darkness is to shine light onto it. Darkness may continue for some amount of time but cannot survive the light of your consciousness.

You will most likely find that the people you spend the most time with like family, friends, coworkers, and so on. Can be the most likely for the mind to become reactive toward. The more familiar the ego is with someone. The more comparisons between ourselves and them can create reactive patterns that strengthen the time-based identity. The identity with resistance. The more we can see this within the moment the mind can't help but react, the less need we have to react. At first, the momentum of old patterns may be too strong to not become lost within a reaction.

A certain alertness is required to be able to question what is currently the most natural way of things for most people at this point in our evolution. However, once we begin to question negative reactive patterns within the present moment, they begin to become more and more diluted with presence. At first, it may not be until after a moment of conflict that we realize what just happened. Yet even this is the beginning of moving beyond the reactions. The more we notice, the earlier we catch them, and the easier we can yield to overcome the direction our mind has naturally gone. This can take many moments of failure, yet these moments of failure are very important for success. While you may have failed on the surface, on a deeper level, you are succeeding in knowing yourself more deeply.

Within this practice, other people become like our helpers in a way. Whether they are ready to understand this or not, the reactive patterns of others can present the most challenging situations to not react towards. As for most people, negative habitual reactive patterns are the most natural way of living for much of the time. Their negativity of various degrees can be very inviting to your reactive patterns. It may be easy to be drawn into disagreements and conflict. While it may be easy to blame them for bringing out these patterns, remember that they are not doing this on purpose. They are likely lost within these reactions more so than you have become. The need to blame or play the victim is also a pattern of ego. You may find yourself playing the victim again and again and again as you slowly but steadily become more easily aware of what the mind is doing.

As you practice presence, many people you encounter will be less present than you are. While they are simply unknowingly strengthening their identification with negativity. They will also provide you with the challenge of practicing presence with someone who is not yet ready to discuss what this means. They may only understand the side of you that is still reacting from time to time. This is what they are looking for most in order to feed the unconscious obsession they are not ready to realize they have. Some people may say you have a positive attitude, but they may only be able to see this in a conceptual way. They likely still need to suffer more before they realize why they don't need to suffer. The helpful challenge is allowing other people to be where they are on their journey without needing to tell them where they should be. This can be a form of resistance that is cleverly hidden within compassion for others. The emotional ego is cunning and will often find opportunities to turn positive intentions into its normalized patterns of resistance.

There is no way of knowing when someone else might become ready to see more deeply within themselves. Yet, once we notice our own patterns, simply yielding and therefore no longer expressing the same energy, we can, in many cases, reduce the amount of suffering others experience around us. They may not know it, they may still be in a negative mental/emotional state, but you no longer add to their negative energy with your own. At some point, they may even ask you how you remain so calm in situations that are obviously very stress-filled for them. They still may or may not understand but will always benefit from the level of peace you have currently found within yourself. At some point, someone around you might become ready because they see something in you that reflects something deep within themselves.

Suppose you carry an expectation to teach everyone, or some of the people in your life, how to find peace within themselves. You will likely cause stress for yourself if they continue to not understand. You can always make suggestions, but if you hold an investment within a certain outcome, it will likely be more stressful than it points towards inner peace. However, if you can simply practice presence throughout your daily routine, it benefits everyone whether or not they are ready to understand why. You are the light this world is seeking. Become more present in this inner light, and you can't help but share it with the world in one way or another through practicing being fully in this one moment. The less we resist what is, the more easily things naturally come together, and the more easily we find simple and balanced solutions to situations.

As you continue to notice resistance patterns, you'll start catching them sooner. You'll start noticing the tenseness within the body more easily. You may begin to notice times when you automatically make certain facial expressions that confirm the stress the mind is unconsciously looking for. It becomes a little easier to let go of the opinion the mind is clinging to and the negative emotions being expressed. As the ability to accept what is becomes your more natural state of Being.

It could, at some point, become easy to start blaming ourselves because we didn't catch the patterns soon enough or shouldn't still be reacting as much as we notice. The mind will try to express familiar resistance patterns wherever it can for the sake of the self-image's survival. Playing the victim is one of the most effective ways the ego gets strengthened. Complaining is a great way to achieve this. Part of this practice is to not take this personally. While every ego is its own unique expression, there are many purposeful personal aspects of our lives with and without ego. Paradoxically, there is nothing truly personal within the practice of moving beyond ego. Only what the mind believes to be personal. The less we take this process personally, the more easily we let go of what is no longer needed.

Ego has been beneficial for potentially hundreds of thousands of years as the strongest of survival techniques for humanity. Effective enough to become the required norm within the human world. With various degrees of insanity within different societies/individuals at different times and places. Perhaps the ego's insanity could simply be viewed as being a byproduct of its success. Until maybe around the time of the two world wars, the first time that our collective insanity reached such a global scale in the ways that it did. When WWII is viewed as a collective experience, beyond the sides that were conceptually created, it becomes easier to see how the energy of the war didn't end, only transformed into the Cold War as two new world powers emerged (the United States and Russia).

This was also around the first time that video footage of the horrific actions of war could create visual documentation, recording proof of just how destructive ego can naturally be on a global scale. Allowing those battles, the aftermath of a Cold War with the threats of nuclear annihilation, wars that followed, as well as many other acts of insanity since then, to be a visual example for all humans to see. Either on the evening news on the internet, as well as those unfortunate enough to live through such unconscious acts of violence to report their experiences on these new platforms. Movies and television shows can be very helpful in painting these pictures as well. Allowing us to ask more deeply on a collective scale, just what is going wrong with our species?

In earlier times the average human could only gather so much evidence pertaining to what a war is really about. Often coming to the conclusion that it was necessary in some way. Whatever necessity there was to fight at the time, it could be easier to overlook the bigger picture that war is madness for the sake of madness. Access to resources can be a major reason for war, for example, but if humans were more conscious, a peaceful compromise of some sort could always be made. Today, we have too much evidence not to have the opportunity to see the collective picture beyond the idea of good and bad, one side against the other. With the internet, it is becoming easier to see the world beyond the notions of us and them.

Until we realize how the mind's need for a conceptualized "other" affects our thinking, as well as our actions, we can only end up repeating the same mistakes again and again, individually and collectively, until these patterns teach us what we need in order to understand that there is no true other. Every individual life form creates its own expression of the natural symbiosis within the collective. The karma of striking down a proclaimed enemy will affect everyone in many ways. Both on the surface levels of life as well as our deepest level of consciousness. If one nation strikes down and or takes over another nation. This will be a giant boost of egoic self-image for many within the invading country, mostly for the egos of those at the top of the hierarchy. However, there will also be many problems they inherit as well as create for themselves within the new societal structures.

Just as important, if we receive no retaliation for acts of madness towards others on the surface level, we will still create more negativity within ourselves while viewing the insanity as a positive that has been beneficial for us.

On the flip side, those who have been invaded will likely hold a grudge for as long as their egos can. Creating a self-image through the negativity and suffering that comes with holding a grudge. Possibly creating a need for the madness of retaliation at some point in time. This is why Jesus taught his followers to turn the other cheek, to forgive them for they know not what they do. It is better to let unconscious people be as they are than to strengthen ego within yourself as well as them. A more conscious society would more likely find a peaceful way, or at least a less violent way, to make necessary changes to situations like these. However, since a nation and its borders are more conceptual than we tend to realize, most of humanity is at about the same level of consciousness, with various degrees of madness, of course. Different nationalized egos can create different forms and levels of madness at different times.

Currently, on both the collective and for most of us, the personal expressions of humanity are continuously creating dysfunction for ourselves and others. This is an important step for collectively creating space for the next phase in our evolution. Being within ego is the only way through the ego stage of consciousness. How much more insanity we need before we can collectively see beyond it is impossible to know. Madness cannot see that it is madness, and we can't put a timeline on the emergence of timelessness. However, within this process, what looks like chaos is actually creating space for something new to come into this world. The madness of ego cannot help but create the reason to shift beyond this level of consciousness. The need to look more deeply within ourselves. It is through the shifting of individuals like you. That is creating this collective shift in our evolving consciousness today, one present moment at a time.

A dream can seem very real, until we awaken from the dream. When the mind begins to resist something, a certain heaviness comes in, a kind of sickening feeling in the gut, which becomes normalized and familiarized in a way that becomes an important part of the mind's identity within most people. With such familiarized emotional reactions, the mind easily sees them as confirmation that its viewpoints are correct, and in most cases, the current situation is indeed a problem. Into adulthood and throughout our lives, habitual resistance patterns continue to strengthen. The conditioned mindset is seen as who we are. Making it increasingly difficult to notice this relationship between the mind and emotions. Unless we begin to see the patterns for what they are from a deeper place within.

Anytime you begin to notice the heaviness of negative emotional patterns, you are becoming more conscious. You are not so absorbed by the mind that you fully identify with these patterns through thoughts and can see them more clearly for what they are. Simply notice how it feels. The mind will try to explain the feelings, but this is not necessary. The ego is very clever and will need to create some mentally symbolic explanation for what is being experienced. The mind always needs a story of some sort for the ways it views the world. Yet the more you can simply be present with negative emotions, the less the mind is able to derive an identity from them. The more the mind tries to explain, the more it will give away its need to hold an identity. Thought doesn't require an identity in order to think, **only an ego needs this.** The less we are trapped in ego, the more useful our mind becomes.

The more you resist negative emotions, the more fuel your inner resistance receives. What they need most is your presence/acceptance of them. The more you can simply be with these patterns without judgment or a need for a future goal of dissolving them, the more the emotional wounds heal themselves within your presence now. The mind and emotions must balance themselves, you are simply the light that allows them to. The more you simply know these patterns for what they are, the less unconscious you can be, and life naturally becomes more peaceful. As more acceptance towards what is now comes into this moment, you will feel increasingly peaceful within the inner body, and more easily, life will flow in a positive direction on the external level.

As we introduce presence into thinking, we create an ability to notice the current train of thoughts from a new angle than the mental identity's ability. Creating more clarity on whether or not the mind is currently attracted to negativity. As soon as we can see how resistance-based a current train of thought might be. New and more acceptance-based thoughts can't help but come in and allow the stream of thinking to become more balanced more infused with a broader and more symbiotic understanding. Compassion for all sides comes in more naturally through a curiosity about how other people may view the situations the mind is thinking about. Every time we are able to be alert enough to notice the mind creating a problem out of a simple situation. It is simply the noticing that yields and overcomes. The simple noticing is what retrains the mind to look at situations in a new way. Turning complex problems back into the simple situations they are.

The mind cannot tame itself, and this situation cannot be solved by the same level of consciousness that defines the situation as a problem. Luckily, it also doesn't have to be as complex as the mind thinks in order to notice the simplicity of this one moment. If there is an awareness of inner spaciousness, it is from a deeper dimension within you that is always peaceful. The ability to notice repetitive reactive patterns presenting themselves is the most important factor. The ability to notice inner spaciousness will increase the more this occurs.

The darkness cannot survive your light. Simply being is what ends the mind's need for restlessness. At some point, you may find yourself grateful for this process, grateful for what the mind has often declared a curse. Life tends to give us what we need in order to become more conscious. We don't always get what we want, but most of the time we find ourselves in situations that challenge us in the ways we need for consciousness to grow. When we resist challenges, they can feed the ego. When we make use of what is here now for the sake of becoming more present, we can only progress in finding inner peace.

There are various degrees of unconsciousness. The more unconscious a human becomes, the more chaotic and even violent they can't help but be. There is no human who ever walked the Earth as a truly evil person. Anyone who has appeared evil is simply lost in the inner dialogue of the mind with which they have identified themselves, likely seeing their intentions as righteous ones. Their thoughts are simply too chaotic to be able to see their actions in a way that a more conscious person would understand as madness. This does not mean no one is responsible for their actions. Only that their actions are lost in an illusion that is just as, if not even more painful within themselves as what is expressed outward. There is no great enemy without extremely toxic and addictive thoughts about the labeled enemy. There are no extremely hate-filled or hurtful actions outwards without the same energy of suffering within.

No matter the level of unconsciousness we encounter within ourselves during the awakening process. We are likely to find heavier degrees of unconsciousness in many of the people around us as we become more present. Giving us the challenges we need in order to move beyond ego. However rough the mind might think these challenging moments might be. However, horribly unconscious people may be viewed, is the illusion that the mind continues to give away as we practice watching and accepting our thoughts and emotions by holding some of our attention within throughout our daily routines. Allowing our reactions to be as they are through allowing other people's reactions to be as they are. Until these situations have shaken the mind up enough times, so to speak, in order to become more aware of ourselves. More aware of inner spaciousness, which is always within us, just covered up by overthinking. It is very subtle, which is why it takes practice being alert and still enough within in order to be more aware of the inner spaciousness. As subtle as it may be, it becomes profoundly peaceful as we gain awareness.

The more we can practice presence while nothing seems to be going wrong, the easier it becomes to notice when the mind declares something is going wrong. Everyone's unique awakening experience requires a different length of time being present. All that is needed is within you now. There is no time involved in this moment, and anyone can become fully awakened at any moment as they practice. There is no way of saying how long or how short your awakening process could be. It is better to understand it could happen fully at any time rather than creating an imagined future date. You will simply awaken fully when you are fully ready to awaken. The real miracle is that it is already happening now.

As mentioned before, like the mental ego, many people's emotional ego wants to thrive on consuming as much negative energy as it feels it needs to, based on personal patterns. Everyone has different levels of this phenomenon, but anyone who is still unconscious of the emotional ego is feeding it to some degree through negative thoughts. There are two phases to these feelings: active and dormant. When the negative emotions are active, they have been able to fully take over our thinking in some way or another. The mind and emotions are now both in this lower level of consciousness than a neutral mindset would be. What we think and do within this timeframe will be heavily filtered by negative energy. When emotions are dormant, the emotional ego is never fully dormant but has been fed the right amount of negative energy from the mind to be able to rest. There are short moments where the emotional body will become active, but not as strongly as a fully active phase. Most people's thoughts will feed the emotional body throughout the day with little random moments of resistance, which can be sustainable, like feeding it a snack here and there. Yet, at some point, it will require a larger meal again in order to hold its ability to thrive. As we awaken out of ego, these patterns may not dissolve instantly because we have become intellectually aware. This is an important aspect, but in order to fully dissolve and move beyond this phenomenon requires a greater physical awareness within the body, which the overthinking mind has been covering up.

For those on the more predominantly negative side of the spectrum. The average day is filled with emotional negativity of various degrees that can be like the hum of a machine. It is always around, so it wouldn't be easily noticeable unless it were to suddenly shut off, which is not likely to happen without realizing these feelings more directly. Some amount of negative emotional frequencies are almost always involved in their thinking. When it is in the dormant phase, the hum of negativity is relatively low. Yet, seemingly at any time, when it has been a long enough timeframe, the negative emotional ego wakes up and becomes very active again. A mental trigger could cause this, it could be the emotions looking for a mental trigger or simply both at the same time.

This occurs in various ways for anyone who still identifies with ego. How often and how strong these patterns are is determined by how unconscious a person is on a regular basis. As an active phase begins, suddenly, they almost become like another person, and in a certain way, they are. They have momentarily become the extremely negative emotional version of their mental persona. This could mean becoming verbally abusive towards themselves and/or others. In some cases, it can mean that they become physically violent towards themselves and/or others. Those who suffer from PTSD may find themselves reliving the past trauma in their mind in some way or another as the emotional patterns are re-establishing themselves. Whatever the expressions might be, they are infused with dense negative reactions.

Then, after consuming enough negativity, the drama goes back down to its dormant level. With a sort of aftertaste left behind for themselves and anyone else who may have joined in the fun. Fun as far as the negative emotional ego is concerned. These patterns will repeat themselves for as long as they can and will likely become stronger as the person gets older unless they become too self-aware to continue losing themselves within the madness. For those who do become more consciously aware, the present moment becomes the most important aspect. These are always echoes of the past that reexpress themselves in the present. The more presence/acceptance we can introduce within an active phase, the less active these patterns can be.

The emotional body wants to feed off the thoughts of negativity but doesn't want to give itself away. Darkness doesn't want to be seen by the light, it wants to stay within darkness. The mental ego is very clever at masking itself for the sake of holding an identity, the emotional ego is very cunning and takes whatever opportunity it can to entice thoughts for more negative food. Both together hold this illusion in place without fully realizing it. Both are lost within their own fog without the ability to move beyond this self-created illusion. Realization requires the introduction of inner light to come in and end the cycles of darkness. When emotions are in their hibernation mode, they make themselves hard to notice as the mind's strength of negativity is reduced as well. Positive emotions become more possible within these times, and negativity can seem completely gone even while a very small amount is in the background in a way.

Depending on the individual's intensity within their active cycle, emotions may always be waiting for the next opportunity to feed on negative thinking. Subtle negative thoughts throughout the day will allow both mind and emotions to feed each other small amounts of negativity until the next full-on active phase begins. It is the emergence of presence that allows us to feel this subtle level of the dormant emotional ego more directly, simply by bringing more awareness into the inner body. By practicing bringing our attention to our breathing throughout the day, the inner body becomes more recognized, inner space becomes more recognized, and the emotions that are denser than inner spaciousness are more easily noticeable within our presence.

It can, at first, be easy to confuse positive emotions with presence. However, positive emotions that are identified as who we are can be like a pengelim that will swing back in the negative direction at some point. The presence will likely help produce positive thoughts and feelings. The difference is knowing what is thought and emotion and what is deeper. As the practice of acceptance continues, inner presence will more easily be recognized as lighter than any emotion. Stillness is another great pointer. All emotions have a sort of movement to them, stillness is deeper than these expressions of energy. Emotions are not good or bad, they are an important part of the human experience. However, like the mind, it is what we identify as who we are that creates dysfunctional behavior within us. An urge or need to feel any certain way is an obsession of ego. The ability to let emotions come and go without being attached is a great sign that you have found a deeper place within yourself.

Presence is like the stern of a ship. Without the stern, the upper decks will flip wildly out of control and likely sink. When we introduce the stern, the ship is now able to balance itself out. Another analogy would be a tree that has grown deep enough roots never needs to fear the wind. These are a couple of analogies out of many, but no analogy is ever exactly what is being pointed toward as words become too conceptual.

Perhaps a closer pointer is, what comes and goes is never who you truly are. Emotional patterns of positive and negative will come and go, with presence, their intensity will take you over less and less.

The deeper peace within you will always be within this moment. As you become more deeply aware of yourself, you become like the light in the room that shines itself onto the darkness. When you are fully taken over by negative emotions, it is very difficult to notice from this state of being fully within darkness. When these energies are calmer, it is a lot easier to practice bringing more presence into what has operated through a lack of presence. At first, this phenomenon will be difficult to notice. It may be very easy for the mind to create an identity around this new practice as well. The mind may tell itself it's doing great one moment, then not well enough the next. Yet the more you simply notice how the inner body feels at any level of emotional activity. The more the light of your presence will transcend the darkness, transforming it into light, into a new level of peace within your life.

The mind may be attracted towards negative emotions, but as you continue to awaken, you may find that thoughts are overall neutral and content with both sides of the coin, positive and negative. Negative emotions may draw thoughts in, but thoughts can be just as attracted to positivity. The mind is very paradoxical in its complexities. This is where thinking about the emotional ego can easily create confusion, quickly creating a mental identity around this process. Ego is very clever and will come back in the back door when it can. Anything that is one way can just as easily become the opposite as far as thinking will assume. That is, the mind will make what it can as complex as it can. Therefore, as we become more aware of this, things become simplified by simply noticing the feelings within the body. Understanding that the mind will sway in any and every direction of good and bad, while both sides will paradoxically fuel each other if we identify with the train of thinking. Let thoughts be as they are, and you will step back from thought and be the watcher of thought from a deeper place.

Our only objective in dissolving emotional ego is simply becoming more aware of emotions, there is no need to mentally label them. The mind will love to narrate this process; it very likely will, and there's nothing wrong with that. The mind loves to narrate anything, and everything it can, but no mental narration is needed anymore once we can see the negative emotional patterns directly for what they are. Simply knowing how they feel, and allowing them to feel as they do within this moment. As well as knowing negative thinking when it becomes lost within these emotions. Acceptance is all that is needed for dissolving patterns of resistance. This, of course, may take some time being present with them, but darkness cannot last long within your light. At some point, this becomes less of a challenge and more of a simple choice to feel lighter within instead of being immersed in heaviness. Letting go of patterns that feel heavy becomes increasingly natural as inner peace becomes increasingly natural.

It is important to note that it's nobody's fault for acting and reacting through ego. It is a part of our species' evolution. People simply act according to their level of consciousness. There's an old Buddhist saying, "If you think you're enlightened, spend some time with your parents." Being present around family can be the most challenging, simply because of how much past they have shared with us, not because of any events from the past necessarily, but the amount of past memories the mind can feed off of. Families also tend to have the most similar emotional reactive patterns, both culturally and genetically. Making them the most challenging people to not express our own resistance patterns around. What is defined as agony by the ego is always an opportunity to notice how this defined agony is only as deep as the surface patterns. This understanding will not transcend our relative's egoic behavior, but will help us from being lost within our own.

This moment is always just simply this moment, no matter what the mind declares. Resolving inner conflict with anyone who involves a past history simply requires the disappearance of time. Not a lack of responsibility for one's actions, but simply a lack of clinging to a past moment in a way that filters how we view this person today. Even if they cannot be present with us, we no longer fuel the madness they may be unconsciously looking for through us. When we no longer cling to any terms of how they should or shouldn't be as a person. There is always peace in the background if/when things get heated on the surface. We simply know from a deeper place within thoughts and emotions come and go but are not fundamentally who we are. We are not the ongoing arguments we might be having with family, we are beginning to notice a deeper level of consciousness that can now see these patterns from a deeper place within. A depth that is at peace with any outcome. The depth of your inner peace is infinite. You never lose it, it can simply be covered up from time to time. Then, as the negative energy passes, you know yourself more deeply again. As this practice moves on, it becomes easier to not be as lost in what isn't real.

By learning for ourselves how to not take things personally, we bring acceptance into a moment the mind is strongly resisting. We start to step back from the thinking mind in a way that was not accessible to us at an earlier time and view situations from a deeper place. We learn for ourselves how to yield and overcome. In turn, those who provoke resistance within us start to become our helpers. At some point, you may find yourself incredibly grateful for the challenges they present. Through their reactive patterns, they kind of chip away layers of ego as they metaphorically bash into the illusory self-image with their own resistance-based identity. This will likely still force us into reaction after reaction for some time. Yet, it is a process the ego must put itself through. Once the awakening process has begun, there is no going back to earlier points of unconsciousness.

The ego cannot help but create suffering for itself. In turn, its patterns of suffering will ultimately lead to its own demise. Metaphorically speaking, ego is being spaghettified as it gets sucked into what looks to it like a black hole that the conceptual self-image has gotten too close to. Any change in behavior will trigger a need to reestablish old patterns that the ego identifies with. In reality, the nothingness of this black hole is the light of truth transforming mental illusion into truth. You are that truth, bringing a new level of awareness to life on this planet. You are the light that ends the darkness within yourself while simultaneously bringing this new light into our world.

To yield and overcome, become still whenever inner resistance is noticed. A certain alertness is needed, but there is nothing to do. Only learning to let this moment be as it is. You may end up in a position where it becomes important to apologize to someone from time to time for your unconscious actions, regardless of their ability to understand your practice of presence. There may be times when you feel a need to help someone else calm down if they are able to. In certain situations, after calming yourself down as you become more still, you will find a more balanced direction to take in the current situation. Simply let the patterns be and watch how they react. They will, in fact, become so obviously repetitive that they can only give themselves away more and more as presence transforms inner resistance into inner peace. Serenity doesn't mean saying yes to everything but simply making decisions without having to resist life. Making decisions from a deeper state of awareness allows a deeper intelligence to come in. Becoming more aware of how many thoughts are not as useful as the mind believes. In many cases, it is the space between our thoughts that can be the most useful.

Some amount of drama is always needed within an egoic relationship. A need to define personal self-identity structures and boundaries will always create a need to defend these conceptual structures while interacting with other people. In an intimate relationship, just the right amount of the right kinds of drama will create more passion on an egoic level, like adding just the right amount of salt to our food. To the ego, just the right amount of shared expressions of suffering make the relationship much more enjoyable. To the extent that many people say the best sex is just after an argument. This could be because some mutual understanding has been worked out. Then again, maybe it's the unconscious thrill of being upset intermingling with the thrills of having sex. Yet, like anything else, the ego is never truly satisfied, or at least not for very long. Ego cannot help but look at relationships with some degree of neediness. Trapped within conditional love, with little understanding or openness towards unconditional love. Which would dissolve the ego identity if embraced enough.

Conditional love, or friendship, always becomes trapped in unconsciously needing something from the other person/people. A strongly defined mental image of what their relationship should be, how the other person/people should be. What others should do for them, how others should act, and so on. Without realizing they are creating the victim role for themselves and strengthening a need for problems. Someone within an ego expression of love will, at some point, become too needy not to create some level of drama. This, of course, may not be right away in a relationship. There is usually a honeymoon period before the drama begins to present itself. Yet, at some point, conditional love will become too demanding to keep from creating problems to some degree.

This will likely cause plenty of arguments and, depending on the level of unconsciousness of those involved, could turn into violent actions. No one can act outside of their level of consciousness. Holding a specific need from a partner in a relationship is likely to create plenty of resistance within ourselves. The more unconscious we are, the more easily this becomes a violent scenario in some way. Ranging from verbal violence to physical violence, depending on the unconscious patterns of those involved. Simply through resistance-based emotional patterns we don't consciously know we are clinging onto. To the unconscious mind, it's always the other person's fault.

If one specific disagreement doesn't do it, the unconscious patterns will find some other reason why the other person is not right, not where/how they should be, or not doing what they should be doing. This could even begin through compassionately wanting to help the other person in some way until the pull toward negativity turns the subject into a need for drama. Seemingly becoming angry because thoughts are in disagreement. However, it is usually more likely because their inner state of unconscious drama needs to find something to argue about, something to feed off of. For some emotional egos, there is nothing more juicy and seemingly fulfilling than fighting with the person we are intimately involved with.

People with similar patterns tend to be attracted to each other. The more conscious we become, the less likely we will stay in a relationship with someone who is less conscious than we are. The mind can only rationalize our behavior or the other person's behavior if we are unconsciously addicted to it. The more deeply we can acknowledge how these unconscious patterns operate, the less we find a need to relive the experiences they create. We may have to leave someone behind, but in these cases, it can be better for all who are involved. Some couples may become more present together. However, it can become unfair for both people if one cannot understand the conscious transformation that is occurring within the other. This may become true with friends as well, but not as likely as within an intimate relationship. There is also always a possibility you may cross paths at a later time with a new set of circumstances.

To experience unconditional love, is to allow any kind of relationship (friends, family, love interest) to be as they are, not without making decisions about the relationship, or suggesting advice for one another. Unconditional love is simply free from clinging to a conditioned state that needs things to go a certain way. Marriage itself is not guaranteed to be an expression of either conditional or unconditional love but is determined by the level of consciousness of those within the relationship. Most marriages experience some degree of both. If a relationship is more time-based than present-based, there will be little to no presence involved. If we are lost in our thoughts and emotions about the other person, much of our time spent with this person will likely be seen through a mental/physical experience, with little to no understanding of the deeper levels of their Being.

If we can't find this depth within ourselves, we will not likely find it within another person. Maybe we find this depth within ourselves, but the other person cannot yet understand. Noticing this could mean that, at some point, it becomes time to leave that particular relationship. It could also simply mean that the other person needs more time to become aware of the presence that you now display before they can go more deeply into presence practices with you. You may even become their guide in a way that could strengthen the relationship. Unfortunately, in some cases, one partner becoming more present can drive the less conscious partner away. As the drama they crave is no longer available through the more conscious individual. It is also possible that while one person becomes more present and the other doesn't, the love/friendship stays strong enough to stay together.

Either way, what becomes more consciously noticed within you loses its unconscious neediness. The same ego patterns may continue for a short time, but consciously noticing is the beginning of the end of conditional relationships. Something unconscious people will benefit from whether they realize this or not. Unconditional love is the ability to see the beauty and joy in allowing things to be just as they are. No longer needing another person to be any other way than they are now. As well as allowing relationships to develop or end if/when they naturally come to this conclusion. The need for someone to be more conscious than they are ready to be is the ego creeping back in, creating drama when it can.

Anytime we notice conditioned patterns within ourselves, we have yielded and have the ability to overcome what is becoming more present. There are no relationships without challenges, but imagine a relationship without an unconscious need for drama. Outside of certain family members, you may find a need to leave some of those who are not present enough. This could also very well mean that you will, at some point, find more conscious people to relate with. Either way, the less conditional our relationship terms are towards intimate relationships, friends, family, co-workers, and so on, the more depth and love we will find within these relationships.

In some cases, especially with family. Allowing the challenges that less conscious people present. Can be extremely helpful in becoming more aware of your own patterns. They can become extremely helpful in realizing that the only reason they can still bring out drama within you is because it is still unconscious to some degree. In most cases, accepting the challenges that they bring with them can be more beneficial than avoiding or resenting these people. This will likely become a balancing act that only you can figure out for yourself what is helpful, and what is too toxic for your current level of development. There are, in some cases, family members who are simply too unconscious to be around for more than a brief moment at a time.

It is important to remember with family that they did not pass down or attempt to pass down unconscious patterns of suffering on purpose. People's level of understanding is based on their level of consciousness. Parents who raise children in any sort of cruel or suffering-filled way didn't do anything cruel on purpose. They raised their children through the logic of the level of consciousness they themselves inhabited and likely inherited. Anyone who inflicts suffering onto others must be lost in the same level of suffering within themselves, which was likely passed down for many generations as the normal way of living. This is not across the board, and humans will acquire patterns of suffering in various ways. Yet, for most families, habitually conditioned patterns of suffering are a continuation of many generations that are not yet ready to face what is within themselves. Confusing their inner suffering with external explanations from the mind. One example would be that it was their parent's fault for how they feel now.

There are definitely plenty of people with strong and balanced parenting skills who raise their children very well. Unfortunately there are also plenty of people who lack those abilities. In many cases, resentment towards the ways in which one generation's parents raised them becomes the psychological karma that fuels the dysfunctional aspects of how they raise their children. In other cases, the ways in which their parents raised them may be a strong enough motivator to do things differently. This might be very helpful for not spreading certain behaviors to the next generation. However, if the resentment has not been fully healed, these patterns are not yet fully present, and suffering will likely present itself in other ways. Until humans are present enough, there is little to no choice involved. The past-based problem will be clung to in various ways. Once humans begin awakening beyond the illusions of time, these time-based problems can be transformed into situations of timelessness.

Now opportunities arise for real choices to be made on how we proceed from here. Now the generations of karma are becoming altered. True healing can occur, and time is no longer a factor. No amount of past can affect the emergence of presence. Generations of abusive patterns can end within a single individual who is ready to move beyond the ego level of consciousness. When resentment for the past is healed, there is no more need to blame family members who couldn't fully understand their actions or karma. There can still be accountability for certain things, but there is no longer a need to hold an unconscious grudge over an unconscious individual.

"Life tends to give us the challenges we need in order to become more conscious when we become ready and willing to accept these challenges. Serenity towards what is allows unconditional love to fully emerge within your life. This allows us to see how all humans, no matter what level of consciousness they currently inhabit, they are an expression of the same one consciousness that is all life, they just may not know it. As if there is an ocean of inner peace within them, only it's covered up by strong mental and emotional patterns that thrive on holding an illusory ego identity together. They are not without this deeper reality, there is simply no way that they can realize how much depth is actually there. They only know the form identity that blocks them from truly knowing themselves. When we find unconditional love within ourselves, we no longer cling to the mind's labels about where others happen to be along this journey toward awakening. Some people still need time to suffer in order to potentially move beyond it. Many will not realize this within their lifetime. Yet collectively, we are all on the same path towards awakening beyond ego.

At the end of any relationship, including the death of someone close, the experience of some level of emotional pain is very likely. As the person we have become so close to is no longer around. The amount of emotional pain we might feel is determined by the amount of resistance toward what we are experiencing. If we can't help but follow thoughts that claim, "This should not be," "This happened too soon," "This should have happened some other way," "I shouldn't have said this," "Why didn't I say that?", or "they should never have done what they did," all thoughts that are based on time. We can expect suffering until the mental subject moves on to something else. Of course, only until this is brought up again. As an unconscious human is not having these thoughts as much as the thoughts are happening to them.

At some point, the mental and emotional pain will not fully go away but will subside to some extent as time moves on. Unconscious patterns will continue to present themselves until we make them conscious enough. The ego needs its suffering, or it wouldn't be able to conceptually define itself. This is how moments of loss can help strengthen ego patterns in most people as they get older. Moments of loss and tragedy can unconsciously be used as tools for the ego to become strengthened or can more consciously become tools for those ready to experience life on deeper levels than thought. Whenever there is emotional pain, there is always some peace within the background, which becomes more noticeable through serenity.

If we accept fully at the end of a relationship that this is what is, and the other person is likely never coming back. There will still be some emotional pain, but acceptance brings a certain beauty along with it. A certain amount of noticeable inner peace. A nightmare is never truly a nightmare when there is enough presence to be able to accept what is. The more we are anchored in presence, the less we are lost in the mind's story about the situation. Which, in turn, allows at least a small amount of inner joy and beauty to come into the life situations that the mind needs to resist. Allowing whatever pain we are experiencing to heal on its own. Without an overactive, resistance-based mind getting in the way.

The more we practice presence when nothing seems to be going wrong, the easier it is to be present when the mind declares something is going wrong. The more we practice simply allowing other people to be as they are, the easier it becomes to accept as well as find beauty and depth within all humans regardless of their current level of consciousness.

One way the ego likes to strengthen itself is through resistance to physical pain. Since life is not without some amount of pain, these experiences can become another opportunity to know the difference between a situation and a negative reaction towards a situation. A surrendered headache is a great example of the conscious act of transforming a huge problem back into a simple situation. The mind obviously wants to avoid the experience of a strong headache. A situation that is never a preferred way to spend the day. This is in no way suggesting that anyone seeks out physical pain for this practice, but when it does come, there are two different ways to experience these situations. One is adding resistance on top of the pain, and the other is simply accepting what already is within this moment. Without enough presence, there is no choice. When there is at least some inner awareness, this opens up the opportunity of not having to add suffering to physical pain. At the very least, there will be a minimal amount of suffering as we practice moving beyond the mind's most natural reactions.

The ego needs to define just how horribly wrong the pain is in order to feed the image of self through playing the victim. The patterns naturally resist whatever painful experience we find within the present moment. Simply noticing these patterns becomes a practice of finding inner peace through surrendering to what is. Once we can notice the mind resisting the pain, we are moving beyond this level of consciousness. Allowing inner peace to become more easily recognized. There is no living without some amount of pain, and there is no getting rid of this fact. It is as it is. However, the real miracle is being able to consciously choose not to follow patterns of suffering when the urge presents itself.

Becoming too rooted within the awakened joy of simply being alive within this moment makes it increasingly difficult to fall back asleep into suffering.

On the surface level, it would appear that life will never leave us alone. If it isn't one thing, it's the other. Always one problem after another, after another. As we become more familiar with our own reactive patterns, a deep inner truth reveals that all these problems become opportunities to move beyond problems. Transforming them back into the simple situations they are. They are always this one step within the present moment, the only moment there ever is. Becoming more aware of the ego requires being shaken by the ego, by its endless defining of problems. Forcing us to become more conscious in order to finally end the endless drama. Through ego, the world seems to be against us both because we're unconsciously seeking drama but also because we still need it for that reason.

This is a wake-up call where our newly noticeable ego patterns need to shake us until they wake us, that is realize what's doing all the shaking from a deeper perspective. In this new environment, every challenge becomes a helpful tool. The more friendly we become towards the situations within the present moment, the more inner balancing we allow to occur on its own. The more we introduce our surrendered presence into the now, the more easily most things work themselves out. What doesn't work itself out becomes more calmly and easily assisted by a peaceful state of consciousness. By shining the light of awareness, the darkness can no longer play the games it used to.

As the present moment continues, we start to see more clearly the two different movements occurring at the same time. The old movement of ego that is always reacting or trying to react towards something. As well as the new movement of deeper awareness, deeper peace, that is simply watching, and sometimes even laughing at the resistance patterns in healthy ways. While the ego still tries and tries to demand the attention it used to get when the game was different. What is steadily but inevitably arising is the ability of the mind to become fully quiet. To be able to go for minutes at times without thinking. Not without the ability of thought, but no longer dragging us around by the endless stream of thinking. Adyashanti, author of "The End of Your World," writes about how destructive the enlightenment process can seem. He says, "It is the complete eradication of everything we imagined to be true." A more gentle way of pointing to this truth is in his question, "What would it be like to no longer be bound by emotions?".

As I write this book, I have not yet eradicated the imagined sense of self to the extent that Adyashanti has. I am currently within the hard-to-conceptually define category of a Human Being who is evolving beyond ego. Also known as a Partial Blossom, although that can become a label for the ego to hide behind. I like to think of it as simply Blossoming, which is an act of blossoming instead of sounding like a stationary object. However, any word is just a word. All that matters is if it can be used to point beyond a conceptual identity. Partial blossom is the part of the human journey where one metaphorical foot is not yet ready to leave the ego, while the other foot can never go back. At this point my mind can still pull my attention away from presence from time to time throughout the day. However, more important are the conscious gaps that continue to widen between thoughts. Allowing the spacious awareness within the body to become increasingly relevant as my personal presence practice continues. Bringing the first metaphorical foot closer and closer to the latter. Allowing inner peace to become more recognizable, as well as more relevant whenever resistance patterns happen to present themselves.

Writing this book was impossible at an earlier time. I had the notion of writing a book on this subject for several years but had to become more conscious of myself before it could all come together. There was always a certain wall of energy that would not allow it to manifest. Too many thoughts I identified with had enough momentum while telling themselves why it couldn't be done, why no one would read it, and many other thoughts of that nature. I had to lose enough of the illusory sense of self before there was enough of an opening to allow something deeper to come through. Once I reached a point of inner stillness where I could write while holding enough attention to my Inner Being, this book started almost writing itself from a deeper place within, as well as through my mind. Both the Human and the Being are necessary for a Human Being to be creative. As far as I can tell, there is an intuitive element to this writing, which feels true through a deep feeling of alert presence that is within me while I write. Now this book seems to be partially writing itself as my mind works along with this intuitive phenomenon within.

This deeper aspect is something anyone can tap into once we become present enough, although it is expressed differently through different people. It is more of a spontaneous guide for incites at unexpected moments than something we can request at any time. Yet, anytime we can have a moment of deep stillness, we tap into this place within us that is the true creative force of the universe. This is how humans are guided wisely when we are ready enough, not by demanding anything but through the simple practice of letting this one moment be as it is. This is not to say we cannot have preferences in order to tap into these creative forces, preferences can be an important ingredient. It simply means if we cling to a specific outcome, ego can likely hinder this ability.

The more we practice turning problems back into simple situations, the easier it is to access this deeper intelligence within that works with the mind but from a deeper place than thought. As I write, the mind is active, but there is always some awareness of inner spaciousness. Allowing the activity to be rooted in presence, rooted in wisdom. In truth, the earlier versions of this book would have been written more from the mind and less from a place of deep presence. This is why, on a deeper level, I knew it couldn't be written until enough inner depth had emerged. The mind simply told itself the ego-based reasons why it wouldn't work, but more deeply, I knew it was a possibility for some later present moment.

As you step into this partially blossomed realm, ego begins giving itself away more and more to the deeper awareness within. What starts as an arduous journey of trying to end unconscious negative reactive patterns mixed with an endless train of thinking. Slowly but steadily becomes a playful dance of transforming illusory problems back into simple situations while finding increasing gaps between thoughts. Allowing the present moment to become infinite, as it has always been. Wherever you are within this transformation, this is where you need to be now. There is no need to wait for some future time with some conceptualized version of being fully awakened. Every time you practice being present now, this moment becomes infinite. Within this moment, you are fully awakened. You just may not fully realize this yet. This simply means that the present moment will continue to give you the challenges you need in order to become more aware of yourself. More naturally able to yield and overcome by simply feeling the presence that has always been within.

Evolving Exercise

That which comes and goes is not fully you but what you are experiencing. There would be no thoughts or emotions without a deeper consciousness for them to express themselves through. Dissolving negative reactive patterns becomes a game of losing yourself again and again within the patterns you are attempting to move beyond. It is very helpful to practice noticing these patterns when they are not strong enough to take you over so that they are easier to step back from when they are.

A powerful exercise would be noticing when other people are reacting unconsciously while you are not engaged in any disagreement or conflict with them. It is more important to notice how they may bring out these patterns within yourself than it is to mentally define, judge, or diagnose their patterns in any way. Practice noticing the subtle inner resistance when near someone else who is experiencing some level of stress. You will likely notice their stress more easily before you become more aware of your own patterns. This requires an ability to notice how their stress may affect you in order to not become absorbed by the energy they are lost within.

In some situations you may find some success in having a conversation to help them become less stressed. At other times you may find they are not in any current state of mind that is willing to become less stressed, less resistant. As their mind has decided this is what needs to be, the rigidness of egoic expressions becomes intensified while they are within this state of mind. Either way, always keep some attention within while encountering other people. Simply noticing how your patterns of stress operate around other people's patterns makes it more difficult for you to become pulled into resistance towards their resistance. There will likely be times when the best solution within the moment is to step away from their energy, not in judgment of them but for your own ability to stay conscious of yourself. This is not a practice of seeking negative people, this will only allow the emotional ego to come in through the back door, seeking its own thrill. This practice is only for when the opportunity arises on its own. There are plenty of unconscious people in the world today, and you will likely encounter plenty of examples of someone whose emotional ego has completely taken over their mind.

In some cases, simply noticing a stranger who's unable to wait in line at the grocery store without becoming heavily irritated for one example. Can be a great exercise for acknowledging both how they feel and how this may make you feel. It is not selfish to use this as an opportunity to become more conscious of yourself. It, in fact, allows you to become more aware of the compassion that exists within us all when we are not lost in negative reactions. You will likely find compassion for where the other person is at. The mind that is resisting them will judge for the sake of its resistance. Simply acknowledging the situation they are lost within is compassion beyond thought. Allowing more compassionate thoughts to come in. Not everyone who helps you in this way is ready to understand how they are helping you. Yet everyone who helps you realize the peace that is within you will benefit in some way by not receiving more resistance from you.

Suppose you find yourself engaged in conflict with someone while you are practicing staying present. This is not a failure but a success in becoming more aware of these patterns. Simply practice noticing what you feel within yourself without needing to mentally label the emotional frequencies in any way. Practice letting inner resistance be exactly as it is, and it becomes weakened through acceptance instead of being fed by more resistance. When you can simply allow the remaining inner struggle to be as it is, that is the end of the inner struggle. No level of inner resistance can survive your full acceptance of it. A tree that has grown deep enough roots never needs to fear the wind.

Partial Blossom

"Those who look outside dream, those who look inside awaken."

-Carl Jung
Author of: Psychology of the Unconscious

Within some speeches at his retreats, Eckhart Tolle will hold his hand out in the shape of a flower bulb. Proceeding to slightly open his fingers outward, only to almost fully reclose and slightly open again and again. Then he might make facial expressions of being physically asleep, barely opening his eyes, only to almost reclose them several times. Giving his audience a visual demonstration of how the mind's momentum can be strong, but the awakening process has its own momentum as well.

Awakening would not likely be a straight line upwards if we were to make a paper chart of our progress. On a short-term scale, it would display patterns of coming out of ego, only to fall back into these old patterns again and again with our progress moving up and down as we repeat the patterns we need to in order to bring more presence into them. Yet, on a larger scale, we would see a steady rise out of ego, becoming more present as the present moment moves on and less able to become as unconscious as we once were. While we may continue to repeat some patterns, there is no going backwards in this process. To know ego for what it is, means we can never be fully lost in it again. As this process continues, there will be moments along this journey where you clearly notice more spacious awareness within you than you did before. As you watch, unconscious patterns become weaker and weaker until they clearly come to an end.

Conceptual pointers may be helpful to some extent but are not the best method. Much like a time based story is not very useful when practicing becoming more present. While we seem to move back and forth like a pendulum that is more or less conscious depending on which side it is swinging towards, we can only move forward as we become more fully where we are now. The conceptual pendulum becomes misleading, and holding a chart of progress in our mind can become meaningless, as there is no way of knowing if a pattern will repeat itself at some later time until it does again. Yet, as we become more awakened, the patterns become weaker in momentum, and it gets easier to simply watch instead of being lost within the reaction. At some point, presence overwhelms old ego patterns, and they will fade away on their own. When this comes, it will be the present moment and can only come from giving these patterns the presence they need in order to heal on their own.

There is no going back. The ego is known for what it is and, therefore, on its way out. The mind may be quick to need to know just how much time this will take being within this partially blossomed realm. However, the greater strength lies within the gratitude towards the fact that it is happening now. At this moment, you are blossoming beyond the ego stage of consciousness. This process will take as long as it needs to take and will only progress forwards from here. The more we can accept the resisting patterns of ego that still present themselves, the more easily they transcend into inner peace.

Ego would love to use a new label like Partial Blossom for its identification with form. Creating for itself a new spiritual ego identity and perhaps spending plenty of time with plenty of thoughts playing around with many other spiritual ideas and ideologies in order to define just where it wants to fit its new identity into the ego spirituality world. Anything that clings to a form based identity is where the mind will naturally go. Anything that is formless is where the new movement of consciousness on our planet wants to go. Not without form, simply more aware of what's always been here, our own living Being. The deeper level of knowing ourselves that is beyond the ego game of wearing conceptual costumes that feel good about themselves one moment and bad about themselves the next.

Partial blossom simply points to the space within our journey that is not in any category, neither the chicken or the egg, while paradoxically within both. Reactive patterns are still operating to some extent, while a deeper understanding is peacefully, and in most cases steadily awakening from the dream. When or where in anyone's life that this parsial blossom phenomenon begins seems impossible to determine. Maybe at the first moment that a deeper realization is found while hearing or reading about this subject. Maybe a little before this point, as the ego reaches a critical density. Maybe it starts in childhood but gets covered by the ego as the mind matures. At this point, natural selection may have put this process into our D.N.A. since intelligence is not limited to what the mind perceives. The deeper intelligence of the natural world may have been way ahead of us in creating this transformation that is breaking news to many humans who are ready enough to find themselves within this category of the collective awakening that is occurring today.

Whatever the conceptual answer of when someone enters into their own Blossoming may be, this might only scratch the surface of the question. A deeper truth would be that Earth is now a partial blossom. Individual humans are at various degrees of asleep & awake, but collectively, Earth as a whole is now within a blossoming phase. Humanity's collective consciousness has started waking up, and there is no going back from here. We are emerging into a new level of consciousness on planet Earth.

Every human that becomes more aware with themselves raises the frequencies of consciousness on the collective scale, making this process easier for more individuals. As larger amounts of the population live in acceptance with the present moment, more cultures around the world will be able to live more consciously, as well as raise their children more consciously, allowing the process of ego to play itself out earlier and earlier in life

Humans, in general,, will slowly but steadily become less violent, less chaotic, and less destructive as resistant-based life transforms into an infinite inner peace that doesn't divide but unites us. Life on this planet has blossomed in many ways, mostly through forms, and the ego is an extreme expression of form-based life, which is why the death of the form is the ego's biggest fear. At this time, it is the awareness of the formless essence of life that natural selection (collective consciousness) wants to blossom into. This is what we experience in between thoughts when a moment of inner spaciousness opens up within our chest, and a deep peace comes in with it. Consciousness is becoming aware of itself more deeply through the expression of the human form.

When large-scale events like Covid-19 appear, the collective ego creates a lot of panic, resisting these situations. What is negative on the surface is usually creating depth on deeper levels. Life is usually most destructive on the surface when a worthwhile change is on its way. While Covid has certainly been destructive, sadly, it has claimed many lives. The dysfunction and chaos created by humanity's collective resistance towards COVID-19 displayed what is truly destructive on Earth at this time: unconscious humanity.

Many people found themselves in arguments over simple details of safety, as well as the resistance towards being advised on how to be safe. Some blamed the media, the pharmaceutical companies, and, of course, the politicians, but they, too, were operating with the same handicaps, on the same level of egoic consciousness. Each viewpoint defended an identification with the chosen side, the chosen opinion they believed to be absolute truth in a dysfunctional way, without the ability to calmly engage in deeper, more beneficial conversations on the topics at hand. As usual with ego, the side I've chosen is righteous and pure, while the other side is wrong, unintelligent, and trying to harm my side, if not everyone. For those who were able to look at every subject without taking sides, the ability to question any piece of evidence became a huge advantage in both seeing the situation more clearly as well as not being as lost in a constant state of fear that was promoted by the collective ego. Not without concern for sickness and death, but not living every moment as if the world was ending like many unconscious people could not help but see the situation as being.

When the mind is resisting the reality of a situation it can't control, especially when death is involved, the mind's need to defend what it believes to be right, becomes just as good. "At least I know I'm right" is how the ego gives itself a small amount of assurance, unconsciously clinging to its suffering while doing so. It is also very likely that Covid-19, like other tragedies of a large scale, has helped many people become more awakened. Things going wrong is usually bad on the surface, but for spiritual growth, things going wrong can become very helpful. To see most people reacting chaotically within an event like this can bring with it a potential awareness of just how dysfunctional people's reactions actually are. Helping someone who is ready enough to become more conscious of their own reactions.

In a more enlightened world, COVID-19 would have helped the majority of the population quickly become reminded of how grateful we are for the things that matter most in our lives. With a calm state of alert presence/acceptance towards the challenges at hand, every question and every idea would have been respected, while in-depth conversations would be welcomed, as well as new understandings of the virus and its effects as time moved on, instead of holding the earliest viewpoints in a war minded like manner against an opposing view. Most people could have learned new information on a regular basis, using the info to help in many ways. Not to claim that I know more details than I do about this virus, but it is likely that mandatory lockdowns, as well as one-size-fits-all treatments, would have instead more likely come to the realization that every individual has different physical needs, and there couldn't be one answer for treating everyone. This more conscious approach would have also helped spiritual growth within the majority of the population to various degrees instead of enhancing the collective ego in the ways that it did.

However, this event happened the way it did because it's the only way it could have happened with the majority of humanity still lost in ego. What they need is to suffer until if/when they come to the point of realizing why they don't need to suffer. Because of this, only a small number of individuals could see the madness of the majority's chaotic reactions for what it was, helping them move beyond their own egos to various degrees. Everyone is exactly where they need to be within the collective awakening, and whether we realize it or not, we are all helping each other become more awakened as humanity moves along its collective blossoming.

"Old friends pass away, new friends appear. It is just like the days.

An old day passes, and a new day arrives. The important thing is to

make it meaningful. A meaningful day, or a meaningful friend."

-Dalai Lama XIV

If our global awakening truly began around the time of intense collective global suffering created by World Wars 1 and 2, the Dalai Lama has been a witness as well as an advocate since shortly after that time. After being forced to leave Tibet in 1959 by the Chinese

Communist Party, he chose to travel the world, spreading the wisdom of the Tibetan Buddhist culture. If it weren't for the tragedy that caused him to leave his homeland, fewer humans would have been exposed to his message of living compassionately towards all beings. Is it possible that on a deeper level, life composed the actions that made him leave? Whether or not the 14th Dalai Lama has definitely made the journey he has been given meaningful.

To the ego, going to war in order to take back a region that was once known as a part of your national identity can appear very meaningful. Yet these types of dysfunctional illusions can only continue to aid in the global awakening process. Just as in the individual, waking up from the global ego is a journey that cannot move backwards, and the dream world can only continue to give itself away from here. The more we collectively expose ourselves to expressions of madness like warfare, the more we can't help but question its role within society. Questioning the level of madness it creates that always far outweighs the supposed benefits.

What is within a dream seems real until we awaken out of the dream. One conceptual example could be when you are asleep at night, and you suddenly notice you can't walk forward as quickly as you are trying to. As you look down, your legs are in motion, but the body is not moving forward as it should be. "Wait a second!" you think, "this only happens in a…" and you wake up in your bedroom. Usually, it takes noticing something many times within many different dreams before not being able to walk or run forward begins to give itself away on this level. This is true when waking up from ego as well. On an individual and global scale, we must unintentionally show ourselves, as well as intentionally remind ourselves that we're lost in ego again and again before we are fully awakened.

The ego's dream world is time-based. Politics can be a perfect example of how the problem is always time-based with what has been done and/or what needs to be done. Rarely, if ever, do we see a politician giving a speech saying, "Isn't this moment wonderful? There's nothing we need to fight against right now!" They may say how wonderful it is that they've beaten the other team momentarily or passed an important Bill. Only to go into what is needed to continue fighting, as the other team will always have a counterstrike of some sort. The need for an "other" will always create this environment. Nevertheless, in its current state, most of our politics can only be an expression of ego. Otherwise, the politicians wouldn't have the amount of attention they believe they need from the general public in order to get work done. In an egoless society, the next election cycle would be less important than the current term. This is a slightly different story for countries without the ability to vote. Yet the less a society collectively supports the current leadership, the less likely this leadership will stay in power.

As these collective expressions of ego become more obvious to an awakening public. More countries will find the freedom to vote, as they should. Politicians in countries that are able to vote will have to change along with the public they try to please. At some point, governments will become unrecognizable from the panicked media frenzy they tend to be on a regular basis in our current world. The type of heavily labelled government policies will become secondary as countries will no longer be run by adolescent-like personalities. The need to have power on an extremely surface level and create an egoic identity through policies will transform into the need to benefit the general public in the simplest ways possible. They no longer need to fight uphill battles with themselves. Politicians will have the ability to find the natural paths of least resistance that benefit all instead of catering to very few. We do see a little of this in today's politicians, but the collective energy is still extremely egoic. However, like all things, this too shall pass.

This doesn't mean that voting is pointless at this time, but it simply points towards how helpful it can be to notice the collective ego in politics while participating at the level we choose to. Without being pulled into the problems they claim. Every defined political problem is an expression of ego needing a problem, and the arguments are just as important. Otherwise simple answers could be found most of the time as it's never really the situation being argued but the political identity. Paying attention to politics for watching this alone can be very helpful in seeing ego for what it is. The display of the collective ego can help the awakening individual more than the individual can help the collective until enough awakening individuals create the momentum needed to flip this around. At some point, there will be too many awakened humans for the ego to thrive in any public office in the ways that it does today. At some point, even the most clever dictator will no longer hold any physiological power over the people within their country.

Many peaceful revolutions may even take place within a short time. This will not occur very easily as long as most people are still lost in ego, fighting to protect an extension of their personal ideological identity. However, once a certain threshold of collective awakening is reached. Simple solutions for government structures which benefit all citizens become inevitable. Clinging to the illusion of power will transform into simply doing the job at hand, and as the Zen saying goes, they will retire when their work is done, allowing the next person to step in.

More important than the projection of an egoless future for humanity is the understanding that we only get there by bringing our awareness to the now. Only when "there" becomes seen as where this moment will be can this transformation occur. It has, in many cases, been through a promise of a utopian future that has motivated humans to go into battle against other humans for a supposedly just cause. Future goals are an important part of growth, but when the imagined future that the mind must get to is resistance-based, this only seems to become fuel for chaos on some scale. Fuel for blaming what seemingly shouldn't be here now instead of fully facing and accepting what is here now. In order to change things in a more conscious way and not repeat the same mistakes over and over again, all we need is presence. Once we no longer cling to a specific outcome, presence allowing a deeper intelligence to come in, and a better society will almost create itself, not without human involvement, but without the ego getting in the way.

It may work out like we imagined, maybe not, maybe something better occurs. Only when we can fully accept things as they are can the mind help manifest the possibilities from a deeper place within. As soon as the victim role comes into play, the cause becomes hijacked by the need to feed the ego. Imagine working for social justice on issues like race, gender, sexual preference, and so on. Without being lost within a need for vengeance over unconscious patterns human ignorance has been lost within. Imagine revolutions within societal structures that truly end resistance towards what is, as opposed to creating a new narrative of resistance.

The collective ego does not end as an individual awakens but is weakened by bringing more presence into this world, which is what all of humanity needs. The world around you will be transformed in some noticeable ways. At the least, those around you will no longer have their egos fueled by you, making their lives a little more peaceful. Some people will not be ready for this and may distance themselves from what they don't understand, or you may need to distance yourself from them. Yet, new people will likely be attracted to your new levels of self-awareness. Friends will come and go, like the days, but a deeper sense of peace will always make the present moment meaningful. You are the light that this world has been waiting for.

Life is never without challenges. The only reason the world is in such chaos today is simply the ego's unconscious need for problems. Yet even that is not a problem, but a challenge that is well needed in order to transform a chaotic world into a balanced one. Even when humanity is fully awakened, there will still be challenges. The difference will be the welcoming nature of these challenges, with a deeper understanding of what they provide. Some challenges may not be as welcomed as others, but all challenges come and go. All situations are as they are, and then life moves on. It's the present moment that will always be here now, and how we choose to live in the now determines whether we live in a state of inner resistance or inner peace.

There are many humans walking the Earth today who could be classified as Frequency Holders. People who are awakened enough that they hold frequencies of higher consciousness. Simply by being present, they help the rising vibrational frequencies of consciousness on Earth. There may be more humans at this time who are still holding the lower frequencies of ego consciousness. Yet this situation is not stagnant, but is a continuous slow but steady shift that is occurring today. Slow and steady for now, but just like the individual awakening, the collective awakening will easily speed up when we have shifted far enough. Holding increasingly stronger levels of these higher frequencies of consciousness.

Someone doesn't have to be fully awakened in order to be a Frequency Holder. The more awakened we are, the stronger our conscious presence influencing the world around us. However, any amount of awareness of awareness itself is this new level of consciousness emerging into our world. The more that awareness of being presents itself in our society, the more rooted society will be within these higher frequencies of consciousness. You are just one person out of several billion on the physical level. However, on a much deeper level, your awakening is equally a part of Earth as a whole, becoming rooted within this new level of consciousness. Any amount of higher frequencies is influencing the whole. Much like any practice of presence, is influencing the individual's overall awareness of self. Any individual who is becoming more present is introducing this phenomenon into the collective.

Frequency Holder can be another term that allows the ego to come back in through the back door. As the mind might praise itself for being a force for good, and perhaps even telling itself it is carrying the weight of humanity's collective level of consciousness on its shoulders. All of this is fine. More unconscious patterns to watch. Yet frequency holding is not a title we can wear on our chest. If we do, we will likely limit the ability with more ego covering it up. Telling people how great we are because we are holding higher frequencies of consciousness may get some people to roll their eyes at you. Maybe it will attract people who hold a spiritual ego identity, who praise you and give you a nice momentary inflation of your mental identity, only to find the deflation effects of ego shortly after, creating fuel for more self-doubting thoughts to come in.

You may even find people who understand what this means on a deeper level than thoughts, and may even help you strengthen your awareness of what it means to be holding these higher frequencies by simply practicing presence together. Either way, it is not who you are, or a thing that you do, but simply a level of consciousness that you are living within. More important than this is simply how present we are within ourselves. When we are present enough, we simply radiate higher frequencies of consciousness. We become a sort of silent example of the next step in human consciousness. Like warm air being introduced to colder air, a physical change does occur.

You will know the difference based on how much more peaceful, lighter, and spacious you feel within. Not everyone around you will understand this, but most will benefit in some ways. As I may have said more than enough times before, at the least, you will be one less stress-filled mind for other people to encounter. At the most, they will feel less stressed when you are around. Feel free to talk about this phenomenon with those who might understand. Just remember that talking or thinking about it doesn't change how present we are. Simply accepting fully what is here now is all that is needed to help change the conscious frequencies on Earth. You may only be one drop in the bucket, yet simultaneously, it is you who is helping this bucket become full.

You may find that animals, like dogs and cats, will notice this phenomenon more easily than many humans do. Not being lost in thinking, animals are much more aware of their sensitivity towards certain energies. A human who is continuously overthinking in a negative way will be far less approachable than someone who is more at peace. The energy of negativity will be recognized instantly and will likely be uninviting to many animals. At the same time, those who lack these negative mental/emotional frequencies will be much more approachable to animals who spend time with humans.

In some cases, even wild animals will feel more welcome in your presence. When we aren't reacting towards a bee, for example, she has little reason to become reactive towards us. When we're not looking at another species through the ego's lens of 'other,' we are far more naturally able to become aware of the oneness that is within them. The one consciousness that is expressing itself through all life forms. They may sense your deeper presence in some way as well. Yet they are also still living in the wild and may not want to be approached. Just as there are many wild species you may not want to approach you.

With domesticated animals, there are some cases where cats or dogs become just as mesmerized by suffering as their human companions do. In these cases, the pet is mimicking the energy of the human/humans they love, and unconsciously becomes addicted to the same negative emotions within themselves as the human does. On one hand, we could say this is a form of animal cruelty. Yet, on a deeper level, this seems to reveal the deeper roles of cats, dogs, and other pets we live with. They keep many humans sane, or at least more sane than many humans would be without them. Through the love of both species for each other, these animals may acquire patterns of suffering as they match the energies of their human companions. Yet we could also say that the love they hold and express far outweighs this minor detail. The healing qualities of love can, in some cases, require this element of sharing each other's suffering in order to find some level of acceptance through this love.

This form of suffering in animals doesn't occur within most relationships with egoic humans. Most pets are free from patterns of ego, at least the human level of ego. Either way, pets tend to raise the conscious vibrations of the humans they love when they are nearby, more so than the human's ego effect on the animals who don't have the complex thoughts to be fully lost. These interspecies relationships bring joy into the lives of all who are involved. This is very likely a part of why humans as a whole have stayed sane enough to reach this point in our evolution. The Earth as a Partial Blossom involves many species. As consciousness wants to reach this next level of awareness on our planet, collective consciousness will evolve through any species it can and will utilize as many different possibilities as it can. Life is singular as much as it is an endless number of individuals. Humans are in the right place at this time to begin shifting to this next level of consciousness, but all life on our planet benefits, and many species could be involved in different ways.

Duality is a subject that the mind cannot seem to help but become lost within, especially in terms of ego identity. As a scientific or mathematical aspect, duality holds a certain importance. When it is a fascination of ego however, it cannot be without some amount of dysfunctional properties. Attachment to one creates cloudiness for both polarities. One issue that an ego can become trapped within is the need for a man to fit the egoic image of a man and a woman to fit the egoic image of a woman. This can brings such resistance towards what doesn't fit this imagined mould that any notion of any other human not acting as they're supposed to brings fear that something is wrong in a way that affects them, even when it has little to nothing to do with the individual that is lost within this resistance. Crossdressing, homosexuality, or transgender can unconsciously seem like an attack on the individual's illusory identity, seeming like a moral value has been violated. In reality, what has really occurred is that the mind has found an outlet for playing the victim role towards a labeled other. Even the notion of a boy playing with a doll designed for girls can create a certain panic within those lost in this strict mindset.

Every individual has a different degree of masculinity and femininity regardless of their gender, and this is true throughout nature. It is necessary for natural selection to express itself through as many different experiences of life as it can in order for living forms to survive and evolve. This means without homosexuality, for example, life would not have evolved as it has. This doesn't mean that everyone must support or strive to support homosexuality in a stress-filled way. Resisting resistance is still resistance. If an issue of resistance towards anything needs to be addressed, the more conscious way, with the least amount of dysfunctional qualities in these actions, is to be completely acceptant of what is. To be enthusiastic towards support, while also finding the ability to accept those who are lost in resistance as well. To forgive them for what they do not understand is to be free from the madness they cannot help but be lost within.

More effective than resisting resistance would be to display your inner acceptance while presenting truthful information, whether someone is ready to see this or not, when it is appropriate. Not needing to resist what is, we more naturally support the variety of expressions of life whether or not they fit within an ideological mould that the collective ego clings to. Including acceptance and compassion for those still lost in ego. Those still identifying with inner resistance are where they need to be for now. When the individual becomes ready, their inner resistance has fulfilled its purpose.

As a social issue, gender, sexuality, and many others can be complicated when trying to educate other people. Education holds a certain importance, especially at this time when resistance towards a labeled "other" is the most natural reaction towards a different expression of life. Yet, as humans evolve, this will become less of an issue. More important is how we choose to approach subjects like these once we have a choice to be at peace within ourselves. The complexities are dropped the moment we accept what is within the now, and it simply is as it is. What could be more beautiful than that? Instead of painfully holding an opinion that claims to know everything and is completely right in a negative way, we get to be curious again and might even learn new things.

In terms of identifying ourselves as homosexual, lesbian, heterosexual, transgender, cross-dresser, and anything else. These are simply the very surface of what it means to be a human being. There is nothing right or wrong about it. However, if we expect to truly find ourselves through these mental labels, we will only find the same egoic side effects as we would within any other mental label. Because no matter what we mentally define, or change about our surface structures in life. If we are not aware enough of ourselves and not conscious enough of mental/emotional patterns, we will always find some level of egoic suffering.

For one example, out of many different possibilities, someone might tell themselves, "This is it! I've always felt like a female even though I was born as a male!" and they may find that life feels better as they make this transition. Yet after a certain honeymoon period, you could call it, they find they are not as happy as they were supposed to be. The old feelings of not being complete seem to still be within them. This doesn't mean that they've made a huge mistake; they are simply doing what life does. Life is a play of forms, they are simply playing through changing forms. The ego may find some other reason why they're still unhappy, but if they are ready enough, the physical transition might help display that our physical appearance is not who we truly are.

In perhaps more extreme physical examples, some people might decide they want to look like a lizard, a cat, or some other favored animal. So they cover their body and face with tattoos. Maybe even get plastic surgery in order to accentuate the physical appearance they have chosen to identify with. A more common example would be people who simply choose to have plastic surgery in order to look younger. There is nothing right or wrong with playing around with our form. It is simply the surface of our expression of life. Our deeper essence is never changed. If we believe we will find ourselves through a surface expression of form, we will only find the very surface of our human experience has changed, and we will likely find some amount of disappointment if we thought that some deeper fulfillment would be found. This is not because we've made some huge mistake. It means that we've seen a simple example of where we cannot fully find ourselves, the mind's identity with form. This example could become a gift that leads us to look more deeply within.

It is, of course, also possible for someone to be fully satisfied with their transition to the opposite gender or from plastic surgery of any kind, with no amount of regret. This, too, could help someone ask deeper questions. They might feel like now that they have been fulfilled in this area in life, the weight of that conceptual subject is off their shoulders, and it can easily show them how much of a surface issue it really is. Allowing them to look more deeply when old, time based patterns continue to present themselves. This could be the same for homosexuality and bisexuality. To come to terms with our sexual preference allows a certain weight to be lifted. Suddenly, allowing us to see that it wasn't as large of a weight as it seemed from the mental identity's perspective in the past. Anytime the mental self image is altered, an opportunity emerges to possibly look more deeply within. No matter how much a surface image is changed, you are still the same on a deeper level than form. Your deepest essence within the present moment is as it is.

In many parts of the world, it is sadly still too dangerous to tell most people if we don't fit into the heterosexual category. In this case, it is more of a safety issue than a personal growth issue to keep this secret. Having to keep this secret is an example of how dysfunctional the ego can be, especially the collective ego. Yet for personal growth, the more important aspect is how we feel about ourselves. To find full acceptance of ourselves is to find full freedom from other people's inner resistance-based ignorance. There is no stronger tool for changing social injustices than being able to accept what is. Bringing acceptance and compassion towards all who are ready to understand, as well as sympathy and compassion for those who are not yet ready.

I am, as I am, one of the most liberating realizations that any human being can find. No two humans are on the same journey, and this is a part of what makes life so magical. Only you get to find what is the best way for you to live. We share our path with others and gain insight from them, but those who are the closest to us are also not on the exact same journey. Sharing our life with others is a great gift, yet only you can find your connection with your source, or collective consciousness, in your own way. It seems extremely personal, and in a certain respect, it is. Yet paradoxically, only ego takes it personally. Unique yes, but we all experience the same basic expression of all life that is shifting towards higher/deeper awareness. To accept fully where we are now is to be fully awakened now, even if not all patterns of inner resistance have fully dissolved yet. Being within the realm of Partial Blossom means one foot is always outside of the ego, even if the other foot is still stuck within. The journey is always the destination, you are exactly where you need to be in order to be awakening now.

The surface, or form expressions in life do not disappear as we become more aware of our depth beyond the form. They simply become less invested in the need to identify as who we are. This becomes a huge relief to no longer have a need to look for ourselves where we cannot truly be found. When there is no longer stress surrounding how the body should look, these surface expressions become more joyous and playful. As there is no right or wrong way to be a human being, there is only this moment now and the choices we make within the present moment. The less we identify with our form, the more likely we find just how much formless depth and peace there is within us. Allowing our surface expressions in life to become more of a playful dance than a stress-filled worry of holding the form together in a conceptually perfect way. No longer a stress-filled need for this or that, to be exact, we realize more deeply that we are the universe experiencing itself through this human body. The body will age, but our deeper essence stays the same as it always has, the present moment experiencing itself within the eternal now.

For as long as humans have existed, comedy has been a source of ego relief. There is a certain wisdom in pointing and laughing at ego/cultural ego in various satirical ways. To break into laughter is to momentarily break out of any lingering ego mood or mindset. Laughter is contagious, and at times, it is difficult not to laugh in certain settings where others are laughing. Something within us wants to keep life lighthearted no matter how much another part of us might want to keep things serious. It was often the Court Jester who would not only entertain the Royal Family but also hold the task of delivering the bad news in some sort of humorous way in order to lighten the reactions that might come.

In many places, the Jester was the only person legally allowed to poke fun at the King or Queen. Anyone else could be punished, possibly even killed, for making fun of royalty. Yet the Jester held a certain importance for keeping the Royal Family from becoming too worked up about any given situation. So the only way to secure this was if there were no rules over whom the Jester made fun of. This worked out well as long as the King or Queen didn't take it too personally.

Stand-up comedians are, in many ways, the Jesters of the modern world. Sometimes delivering material that is purely lighthearted, with a quick set-up, punchline, setup, and punchline format. Comedians can also use laughter as a way to explore deeper issues within societal structures. Bringing in deeper and sometimes darker subject matters while delivering punchlines at various points within their storytelling. Narrating subjects in ways that can make us laugh at things that the mind would normally want to take too seriously to be joyful with. When it's executed well, comedic entertainers of all art forms become shaman-like or spiritual healers in a sense. Bringing us to a certain place of inner openness that can allow deeper messages of wisdom to be more easily understood. Sometimes even breaking through certain shells of egoic ideology through our own laughter. One could even call them teachers, as they continuously teach us how life never needs to be taken too seriously.

Comedians are healers and teachers as long as we don't take something too personally, triggering a reactive pattern for the ego to come back in. Suddenly, what was extremely funny is now extremely uncalled for. This, too, can be a great opportunity to become more aware of how the mind is always waiting for that next line to be crossed. Waiting for the next seemingly important reaction to become activated. It is only when we take things too personally that the ego can reestablish its personal brand of suffering. When we notice this, we have something else to laugh at: our own imagined sense of self. Laughter can dissolve any current patterns of ego as long as it is rooted in self-awareness and not rooted in egoic energy. However, when we laugh at something or someone in a way that puts them down for the sake of momentarily inflating our own imagined self-worth, we can only fuel our ego. When we laugh at ego, we are liberated from it, at least momentarily.

Issues about race, sex, gender, politics, war, and so on can be very touchy subjects. Especially in a society that clings to the idea of my thoughts against the "others" thoughts as much as many people do today. Yet, when we don't take an art form like stand-up comedy too seriously, joking about these issues can have the potential to help people see just how similar we all are. While life should be serious in a certain sense within certain situations, we don't have to be lost in our egoic reactions of seriousness. Comedy can sometimes point out our differences in a way that seems hurtful when taken out of context. Yet from another viewpoint, poking fun at anything and anyone, for most professional comedy, is meant to do so in a way that shows how connected we all are through laughter. All humans are in the same conscious state while laughing; different thoughts may occur within our consciousness, but we all experience the same joy-filled state of awareness. No matter what our physical form or our cultural heritage is, we all laugh at a good joke. Laughter can be the liberator of mental heaviness.

Life can be serious if we choose to see it that way. Not that it is a true choice for those lost in suffering. Yet, for those who are able to make it a choice, life can also be playful and humorous. Even the greatest tragedies have some amount of peace in the background when we know ourselves deeply enough. The surface

241

aspects of any situation can be respectfully taken seriously when needed. While on a much deeper level, noticing some amount of joy from simply being alive. If we learn to laugh in appropriate ways, and sometimes even inappropriate ways, laughter can shine a light on the darkest aspects of our lives as long as we have the wisdom to not take life too seriously. Comedy becomes a part of our awakening as it becomes a tool for us to remember that life is meant to be joyous. The old ego patterns of bleakness become more easily laughed at when we don't take them too seriously. To laugh at ourselves is to know ourselves more deeply than any current patterns of seriousness. Laughing at others becomes egoless as well when we can laugh at ourselves in this way. Allowing comedy to bring more light into our world as it loses any purpose of inflating old ego patterns.

Before modern inventions that could record sound, people could only enjoy music when there were instruments around, along with people who knew how to play them. The euphoric experience of listening to music that is performed by professionals was limited to being in the right place at the right time. With the inventions that came from harnessing electricity, music has transformed into something we can enjoy at any time we wish. Which song do you want to hear now? Play it again and again if you want. The ego of course, will identify itself with the music we enjoy, much like a sports fan will identify with the team they support. Reestablishing its identity with this song, this album, this group/artist, this genre, and so on. Holding a part of the overall identity together with the conceptually labeled musical identity. While this holds the ego together, the depth that music brings into our lives makes this one of the easiest categories of ego identity to dissolve, at least temporarily.

The music ego, or imprinting an amount of self-identity onto the music we enjoy, is most likely one of the first elements of ego to fully dissolve by the light of our simple noticing the way the mind thinks about and clings to the music we identify with. This could mean we no longer listen to certain music. We may find new sounds we weren't into before. This may even strengthen our enjoyment of certain music. I've found many songs and albums that artistically point out ego from an egoic perspective that now have more depth than they did before I was aware of ego within myself. Something about the way music relaxes the mind or alters the train of thinking allows even the most egoic or suffering-based lyrics to be realized from a deeper angle. Any layer of ego that is dissolved, creates an example that helps us dissolve more layers. In some cases, a song that once encouraged an identity with suffering can be transformed into encouraging deeper awareness of suffering. Bringing the light of our awareness into a dark subject matter, along with the rhythm we enjoy, helps us find deeper peace within where there used to only be reestablishing suffering. Not every song/artist will do this, and some will simply feel like there is just too much ego as we become more aware of ourselves.

Music can strengthen the ego, point out the ego, as well as point beyond the ego. More so, music can allow space to come into this moment, allowing at least some thoughts to subside momentarily. Allowing deeper peace and sometimes deeper wisdom to come in. Especially music that is not filled with lyrics (thoughts) or has periods without lyrics in the songs. Silencing the mind is the goal of being free from endless thinking. Being in complete silence is very powerful for this, but somehow, music is another way of listening to the infinite depth that is within us. I couldn't have explained it this way at the time, but I feel that when I was growing up, music held the first spiritual teachings I experienced. Both through messages of wisdom within certain lyrics, but even more so through the pointing towards depth through the rhythms of sounds.

The therapeutic qualities that music provides can be powerful in many ways. A great musician displays a certain depth through their music that goes beyond the artist's normal levels of self reflection. Just where exactly does an artist's creativity come from? Any artist who creates something new seems to pull it out of thin air, from someplace deeper than their form or mind. The music created throughout the twentieth century seemed to have had a collective depth behind it that went far deeper than the individual artists. As if a collective expression of humanity was being displayed, as well as amplified through this new experience of recorded music, and was listened to by larger audiences than before the invention of the radio. Television may have increased this collective creativity experience as well.

The depth that music is able to point towards was, in many ways, for the first time, amplified to unmeasurable lengths and depths. Then by the end of the twentieth century, this phenomenon decreased to some extent, and in many ways, music itself seemed to become less influenced by this mysterious expression of depth within collective humanity. Many would say that music today simply is not as good as it was, and may have many reasons why. Maybe it was more necessary during different times in the twentieth century for great music to keep people sane, as I've been told by many people of the Baby Boomer generation. Yet today, there are more types of popular music than there were, and the internet can make it easier for each individual or group to become buried within all the different selections.

Perhaps great music with that same amount of depth and ability to speak beyond the artist is still all around us. Yet, simply watered down by more popular, that is, more corporate-embraced music that is, in many cases, much more about the looks of the artist and a familiar sound as opposed to anything experimental. Creating less ability to find more depth within their artistic expressions than the producers trying to sell a product can understand. Perhaps I am limited in understanding today's music and the depths it may hold because it speaks to a different audience than myself. This could be very true, but today's music still seems to lack a collective depth that earlier music seemed to harness. It still can be very powerful in many ways, just missing that deeper element. Maybe another pop cultural wave of music with great collective depth is just around the corner. Artistic patterns like these cannot help but repeat themselves, pointing out ego and some amount of artistic understanding of shifting beyond it. Regardless of the specific theories I've just stated, all music can play a vital role in our evolution beyond ego.

A great example of how music created by a single artist or a handful of artists can speak on a collective scale. Is within the post-World War Two generations during the second half of the twentieth century, especially during the collective energies surrounding the 1960s Civil Rights movement and protests against the Vietnam War movements in the United States. Different genres of music that were simplistic in many ways in the late 1950s quickly became much deeper and more complex in many ways by the end of the next decade. Many groups from Britain seemed to accelerate this expansion of creativity, for example. Seemingly out of nowhere, different areas of the world would have music scenes with many artists competing to be as great as they could, collectively raising the bar for collective creativity.

One example could be comparing Buddy Holly with Jimi Hendrix. Both were great artists with great sounds. Yet the depths and complexities of the Jimi Hendrix Experience seem to be otherworldly in comparison to late 1950s Rock & Roll. It is amazing to ponder how popular music made such a leap within such a short amount of time. This shift was, in many ways, a direct reflection of the world at that time. Post World War Two was a time of opportunity to shift how we live as a species through the reflection of how current societal structures inevitably create continuous war on many levels. Unfortunately, and very understandably, this was not understood in this way by many people at the time. When great changes are being created, it often looks more like chaos for the sake of chaos, but within the aftermath of two world wars, the next couple of generations were almost guaranteed to ask themselves, "Wait a second, why do we live so violently?", to at least some degree.

Suddenly, within this timeframe, collective art forms were reflecting the current state of humanity in ways that can be difficult to understand just how so many young people carried such deep wisdom that they very naturally expressed through their art. Yet they simply tapped into a deeper place through their artistic expressions, a deeper collective wisdom that is within us all. It is very likely that these were expressions of the first movement towards collective awakening brought on by the Second World War's extremities of suffering. Not a full awakening, as we were far from ready for that at the time, but we had become more than ready to begin the process of collectively moving to the next level of consciousness, one of the first baby steps, if you will. An early collective wave of deeper questioning rolled through the cultural expressions of the time, with various degrees of understanding within any individual.

Most of these artists were still very ego driven, and becoming extremely famous at a young age doesn't help that. Yet for those who found joy in practicing with their instruments for several hours every day. A much deeper collective expression of music during that time was able to come out of the individual's unique talents. These musicians became very shaman-like as they pointed the world towards deeper reflections of ourselves and the collective societal structures we live in. In many cases, pointing out expressions of the ego from an artistic point of view. With a melody that allows the listener's mind to go into a state that is not the same as being deeply lost in thought. Allowing more depth to come in, whether it's noticed or not. The fact that these songs were recorded means that those artistic expressions live on today.

This is not reduced to the phenomenon of music within this timeframe but is, in fact, what most art does. It goes beyond the talents of the human, creating a glimpse of the world and ourselves from a deeper level. Art has benefited our species in many ways throughout history and is now beginning to point us more and more towards an awakened world. Much of our art today points out how chaotic the human world is from a heavily filtered egoic perspective. Yet without artistic reflections towards the flaws in our dream world, it might take us much longer to awaken from the ego's collective dream. There are also growing amounts of art today that cannot help but point beyond ego, some more consciously than others. Alex Grey's work is one example of art pointing beyond the world of form towards the spiritual dimension that is within us.

Another phenomenon that can contribute towards our individual and collective awakening is the use of psychedelics. However, they are not necessarily needed for any individual awakening process. As well as, it is also important to mention that these experiences are not for everyone, and **especially** if used for the first time, should be with the supervision of an expert, or guide. I cannot say what is appropriate for any individual's path. It is always our own decision to make towards what may or may not be helpful for ourselves. Psychedelics can have the ability to momentarily pull the mind away from the ego and create a space for new understandings to come in. Humans have used psychedelics throughout ancient times for these reasons. Until certain cultural shifts at different times and places began to view them as only harmful and, in some cases, evil. Unfortunately they can also be misused when misunderstood. Especially when used purely for recreational purposes, overlooking the values of inner reflection they potentially present.

During the counter cultures within the 1960s and 70s. Many people described finding a certain inner harmony with life or a newly found depth within themselves while experimenting with LSD, Psilocybin,

Mesculin, and other psychedelics. Unfortunately, the lack of understanding towards these experiences would, for some, lead to the notion that these substances were the answer to all of life's problems. Instead of being able to realize that they can only sometimes help with certain challenges in life but are not a key to the universe, as some would say. As much as they may be able to teach some of us about ourselves. Simply being present with ourselves is the key ingredient for deeper self-awareness. Whether we choose to use these substances or not.

For those who overused or used within improper settings, these substances may have produced more negative effects than beneficial. The countercultures of the mid-twentieth century had been long since cut off from the deeper understanding of what these substances can do as a tool for self-awareness. Unfortunately, they were mostly viewed as party drugs. There were likely very few people who didn't use them in these unconscious ways. However, even misunderstanding the tool that psychedelics can be, they can still display some beneficial results. The question is how well the individual notices what they experience.

In the early 1970s, the drug war started in the United States, and almost all information about psychedelics, as well as other psychoactive substances, became misinformation. Strengthening this propaganda trend that exists in many other countries around the world. Claiming they are only harmful and have no medicinal and certainly no spiritual growth value. Today, this is turning around as laws are slowly being lifted in certain places. Studies of therapeutic properties have proven that they can help some people with PTSD, depression, and addiction towards other substances, as well as having other self-reflection properties.

When substances are used as a form of escape, they cannot help but create more unconscious problems in life. When used medicinally, it can be a different situation. There are some therapeutic as well as medicinal qualities within substances like cannabis, for example. One reason is that the THC slows down the normal momentum of thinking while relaxing the mind. It can also be helpful for creativity at times. Unfortunately, it is not without some degree of unconsciousness. Especially at the opposite end of the high, when the after-effects are strongest. What goes up must come down.

Alcohol use can be therapeutic as a stress reliever as well. Unfortunately, there are some people that become more prone to negative ego behavior the more they drink. One or two, and they may be okay, but one too many, and they may seem like the beast has been let out of the cage. As if they have become a whole other person because, in a way, they have. The intoxicating effects have made them less conscious. Making them more prone to unconscious ego patterns. In some cases, violent behavior becomes more prevalent. It is possible for some substances like cannabis and alcohol to offer some amount of deeper perspective on life, a form of

251

stress relief, as well as a tool for creativity at certain times, just with the trade-off of not being helpful in a fully conscious way. It is likely that as humanity's consciousness rises, there will be less of a need for mind-altering substances. As they become no longer necessary for finding moments of peace, or tapping into the depth we are at this point always able to find within us.

Today, psychedelics are being used by more people for their therapeutic purposes, with the knowledge of appropriate settings and doses. They can help many people with psychological, as well as spiritual aspects. Not a therapy for everyone, but one that can benefit many, and is contributing in some ways towards the blossoming of humanity into higher consciousness. Scientific discoveries have found that psychedelics have been used in these ways since at least the time of Neanderthals and may span back to our earliest hominin ancestors only cut off from our cultures in recent history.

I cannot tell anyone what is useful in their personal awakening journey. However, for those who feel it is helpful, psilocybin, as well as other natural psychedelics, seem to help us tap into our deeper level of consciousness in a way that lets us realize things that our normal train of thinking will not allow. There is also evidence that they may benefit the brain's performance by strengthening neurotransmitter connections. Yet much more scientific research is needed in order to answer many questions about the physical benefits these therapies can have.

While these substances can be temporarily helpful and, in many cases, euphoric, the deeper therapy is in the aftermath. Recognizing the deep realizations our own inner wisdom provided, and utilizing them afterwards in our daily situations. It can be especially helpful to bring the practice of presence into the whole experience, both during and after. Any inner realization we find while the mind is experiencing these substances is the wisdom that is always within us. Psychedelics might temporarily point out, yet psychedelics are not necessarily needed and are not for everyone.

Practicing presence is what allows a deeper perspective to come in, regardless of the current situation. A bad trip, as it can be called, only occurs when we resist what we are experiencing. True wisdom is neither bad nor good, as it transcends both. We may find a deep truth to be temporarily painful to face, yet the more we can accept what is within this moment, the more peace we can find within what is. Egoic resistance patterns cannot survive our full acceptance of them.

On the collective scale, psychedelic plants and fungi have been helping consciousness evolve on Earth for longer than humans have lived on it. They may have started out producing these molecules that affect our minds and bodies in order to defend themselves, as well as their offspring. Yet somehow, life on a larger scale has found a use for them in this other way, much like all the

253

other natural medicines on Earth. Maybe humans and other animals found them by accident. Maybe a deeper intelligence that is not bound by forms was waiting for all the different natural medicines to be found by accident. Maybe it's just as simple as letting nature take its course, and life naturally finds a way.

Many of our medicines in the modern world are created with the intention of keeping the customer coming back for more, as opposed to trying to cure the patient. When the people working in these fields become more present, the quality of the present moment will serpace the quantity of a supposed future. All work environments, as well as what they produce, will be transformed by bringing presence into the workplace.

In today's world, the ideology behind consumerism is to get people to buy the next product as soon as possible. A means to an end that "makes the world go round", as they say. A means to an end that always creates a grind to strive towards the next moment, the next deadline. Always promising the future, the next product will be even better. Making it impossible to produce a product that brings quality to all life on Earth at this moment.

In an awakened world, technology will have no need to constantly advance. As life becomes simpler, technological advancements will continue but become secondary to the quality of all life on Earth. Most of our needs to live are very simple. Our economy could be just as simple, depending on how we choose to live. If humans could make the present moment more important than the future, there would be no future utopia to strive for. The reality of the deep peace that is within us all in the present would be far greater than a mental yearning towards the future. As well as clinging to a past that can never be again.

These understandings may not be seen yet on a global scale. Nevertheless, you can make this wisdom true within yourself anytime you bring your attention to this one moment. With full acceptance of what is in the now, the world is beautiful just as it is. There may be some temporary ugliness on the surface, but that, too, has a certain beauty to it when we let it be as it is. The darkness is never as dark when we shine our inner light of acceptance onto it. You are the awakening light of planet Earth. However blossomed or partially bloomed you may be, your presence is the light of the awakening world. A light that is timeless and has always been.

"I am the sweet fragrance of the Earth, the light of all beings."

-The Bhagavad Gita

Today, we are at the beginning of a global shift towards the next level of consciousness for our species. Like most individuals, it will likely be a slow and steady shift for our collective awakening. However, as we reach certain thresholds, it is also likely that large fluctuations occur at various points along the way. Large percentages of people may become more conscious within short timeframes. There is no way to predict just how long this shift will take because the present moment is all that is needed. When collective individuals within our society bring presence into their daily lives, the peace within inner serenity will overwhelm the need for societal structures to always be restless towards the present moment. Like our individual journey, a collective shift in consciousness will take as long as it needs to take.

Throughout human history, there has always been a very small percentage of fully awakened individuals, as well as those who are partially awakened. Today there are very likely more than this average, yet it is perhaps impossible to know what the correct percentages are within these beginning stages of collective awakening. With the numbers rising, I am hesitant to write any estimates down. Yet, roughly speaking, there is likely under a quarter of the population that is both fully awakened and somewhere within the process of becoming more conscious. It is impossible to know the timeframe of how this shift occurs because timelessness is what is most needed. Yet, like our individual shift, there is no going back to any earlier stage. Once this shift has begun, the cat's out of the bag, and our collective shift can only move forward from here. Provided that no societal tragedy occurs on a large enough scale. Yet even that would only delay the inevitability of our evolution.

Many people today seem to speak/think about deeper reflections with only one limit. The mind's thoughts come across the barrier of identification with thought. Meaning there are some awakening qualities, but they still fully identify as the thinker. As they ponder deep subjects, they always reach a limit of how deep they can realize within themselves. They likely possess many qualities like a decent level of compassion, empathy, and an aim for all people to live in peace. Making them partially blossom in some sense. They simply haven't come to the full enough realization that they are not having thoughts, but their thoughts are, much of the time, happening to them. Repeating continuously in an unconscious way as they identify with what is happening to them.

They may have found a deep empathy and compassion, but not without an unconscious need to also judge in some way, at some level. Their biggest difference is that a conditioned way of thinking cannot produce unconditional Being or unconditional love. Experiencing compassion conditionally means that it is absorbed by some degree of the mind's need to feel good and bad in some way. This is still a great sign that many people are becoming ready to look deeper within. Compassion is always able to point beyond conditioned patterns to some degree.

Critical thinking skills are also a very important aspect for anyone to not become trapped within, or identify with any specific opinion. The ability to question any thought we have will allow some amount of seeing beyond the mind. Those who possess these qualities are, in some respects, moving beyond ego. Yet unless someone reaches a threshold of some kind within themselves that creates a reason to shift beyond identification with the overthinking mind, we as humans can only see our thoughts as who we are. We can be great at managing how we think but cannot keep ourselves from overthinking. Cannot help but be absorbed by the mind to some degree.

So far, it has been those who suffer most from the thoughts they have who have the highest likelihood of disidentifying with ego. As we continue to collectively shift, this will become less necessary, and more people will be able to move beyond ego with very little suffering involved. Society as a whole will shift from a suffering-based society with all the self-punishment, as well as punishment of others based expressions that an egoic society needs. Into a more compassionate, more peaceful, and less dysfunctional society. There will likely still be regulations, laws, governments, and so on. Yet public policies of all kinds will become less and less likely to stumble over themselves with the dysfunctional qualities of ego.

The prison system, for one example, will likely be transformed from a punishment-based system that creates wealth through fines and puts people behind bars into a structure of mental health and rehabilitation. It will even likely spend more time on crime prevention strategies as opposed to fighting an endless battle against ideologically labeled criminals. There will likely still be facilities that hold people for a certain amount of time, yet they will probably look much more like rehabilitation centers as opposed to putting people in cages for the egoic understanding of punishment-based justice.

Justice from an ego's perspective is taking something that is defined as wrong and balancing it by making it right. Unfortunately, punishment is rarely effective without dysfunctional consequences. Not that it's never effective, but it tends to reestablish the negative as opposed to truly finding balance. Rehabilitation beyond egoic understanding requires seeing beyond good and bad, right and wrong. When people are no longer villainized, the motivation to act like villains fades away on its own. When the people within the prison system are more conscious, they will more easily point towards inner peace as opposed to strengthening inner resistance patterns.

There are many complexities towards crime and crime prevention. Yet, when we look for the paths of least resistance through empathy, simple solutions will present themselves more easily. When the prison system is no longer looking for an imagined justice against a labeled evil, human beings will be given the necessary help they need in order to be less dysfunctional in their lives. One side against another side will turn into finding what is best for all involved within any particular situation. It is hard to know just how this system will work because it will be unrecognizable from today's Judicial Systems.

Governments will likely transform from a circus of egoic personas clinging to illusory power structures into a humbled bureaucratic system that simply completes the tasks at hand and retires when the work is done. Allowing the next group of candidates to come in and do the same. When the ego is out of the way, things can actually get done, and progression is no longer a struggle to achieve. No longer one step forward and two steps back as ideological viewpoints fight against each other. Awakened humans will be able to simply do the work that is needed at any given present moment. There will be less and less of an excuse for war as a pursuit towards peace or for the sake of defeating a labeled enemy. Nationalism as we know it today will fade away. We will certainly still have different conceptually defined areas of the world with different cultural backgrounds that are honored in many ways. Yet the strict ideology of nations will dissolve as they become less necessary for serving the purpose of ego inflation towards an imagined "other". Grudges between nations and cultures come to an end very easily when we are content with our own state of being.

A peaceful world is the easiest thing to create when we find peace within ourselves. At this point, we as humans can fulfill our fate of becoming the keepers of this world and bringing true order to it, as opposed to always bringing chaos and destruction. We can make Earth a better place for all species to live while terraforming it back into the rich environment it was before the industrialized learning curve brought us to the environmental challenges we face today. Life on Earth can thrive with a higher level of consciousness, paving the way for more and more joyful ways of living. This doesn't mean it will be simple or without challenges. Life always needs challenges of some degree in order to thrive, and grow. This simply means that we will more naturally face the challenges at hand when we are no longer unconsciously pushing the present moment away.

Within a fully awakened world, the monetary system as we know it may not survive this shift. It is hard to say, as it too will be unrecognizable when it is no longer based on survival of the most ruthless and clever. More conscious wealthy humans will become more charitable through the simple realization that their income does not bring them happiness. Their income is not who they are but an extension of the egoic persona if it's something that they cling to. There could likely still be a system of purchasing materials with funds, yet the need to struggle in order to obtain food, housing, clothing, education, and so on will likely fade away as fewer people are trying to extort as much money as they can for selfish reasons.

Money is not good or bad but simply a tool used for trading purposes. Money is nowhere near as important as the ego makes it out to be. There is an importance, but it is only the ego that worries about it, only the ego that clings to it, only the ego that turns it into a form of obsession and suffering for ourselves and others. Some sort of monetary system will likely exist in a more conscious world, but the level of importance will no longer be of life or death. There's no way of knowing many details, but when there are no longer strongly defined others, the survival and well-being of everyone becomes understood as enriching everyone's lives. A world with some amount of shared financial wealth through charities or investments brings wealth of love, education, creativity, innovation, health, and many more positive aspects to all individuals. Hoarding money can't help but create the opposite. This is not saying that everyone will have to share their finances, just that more empathy and unconditional compassion will make government programs like this more likely.

Would this make for less production, or would this simply make for better living standards? Will only some people find more beneficial careers in life because they aren't trapped in a financial grind they can't escape, while more people simply become lazy with less of a need to work if everything they need is paid for? These are questions that can only be answered as we become a more conscious society. As we become less a species of doing and more of a species of Being, we will find a balance within our societal structures that doesn't move too far in one direction or the other. As the wisdom of Being doesn't cancel out the wisdom of doing, only balances out our priorities with both.

Consciousness wants to evolve on our planet. It may evolve through you, and/or it may evolve through others. Either way, consciousness has evolved to the point of this ego stage and will evolve beyond it. There will come a time when this becomes more natural for our species, and this next level of consciousness blossoms through large numbers of humans. What human society will look like will be unrecognizable in many ways, yet daily life will be essentially the same. People will simply get up in the morning, complete the daily routine, and go to bed at night. The largest differences are a lack of endless mental chatter, a lack of endless problem-making, and an infinite joy of being fully in the present moment. This is how the global shift starts and ends with the individual, and your awakening is simultaneously the world's awakening.

However awakened, or partially awakened, you might be within this moment, the conceptual rating of just how awakened is not as relevant as the fact that you are awakening. Even the smallest amount of presence in this world is a miracle to be cherished. The mind can easily get stuck on details as to why you are not as awakened as you could or should be. Thoughts may love to repeatedly tell the story of how far you have come, as well as how much farther you are not. Yet every moment you are able to bring your attention to this one moment, it becomes irrelevant just how awakened you may or may not be. Partially blossomed is only partial on a conceptual scale. The deeper truth is that you are awakening. You are fully awakened, while paradoxically, the ego needs more time to become timeless.

Thinking may still have a certain momentum to it. However, the simple ability to notice this is from a fully present place within. Whenever there is an anxiety-filled need to be more awakened than you are right now, notice this: be the silent watcher. Let the mind's resistance patterns help you become the watcher of the mind, as opposed to the belief that the mind is all that you are. The less you follow thoughts of separateness, the more naturally you will find the deeper knowing that we are all one within the dimension of Being, not through a conceptual knowing but in a way that the mind cannot know on its own without your presence.

Let yourself be as awakened as you can be today, and the awakening process can only continue. The more you allow yourself to enjoy the ride of awakening where it is now, the easier it is to know that you are exactly where you need to be right now. What you need in order to become more present will become more relevant when the time comes. Use what is around you or what challenges life brings to become more deeply aware of yourself. The more you accept the present moment as if it is what you want, the easier the peace within you becomes recognized. The miracle isn't reduced to an imagined finish line that is lost within a conceptualized future. The real miracle is that you have found the path of awakening now.

Endless numbers of generations have helped you get to this point of readiness. The place where suffering through conceptual time ends, and the peace within timelessness truly begins. Now is the blossoming of collective consciousness; your self-awareness is the peace that our world has been waiting for. Not the mental labeling, or ego clinging to this, but the deeper experiencing of your true self shifting out of collective ego into eternal peace that needs no label to be as it is.

Evolving Exercise

When the mind is certain that a goal of being fully awakened needs to be met, we can never be there. Yet when we can accept and be grateful for how far we have come, we've already found the most important element. Noticing when the story of time is claiming a need for a future in order to be free from the past is the strongest practice towards allowing yourself to be where you are most needed. Hear now, no matter how much momentum thoughts may still have. Any moment you can accept where you are along this inner journey, you are where you need to be, and the collective awakening process is exactly where it needs to be. There is no place to go, and nothing to change into, only the practice of noticing that it is the looking that is becoming increasingly aware of the difference between a thought, and who we are.

WATCHING THE RIVER'S FLOW

"Choosing the path of wisdom, become aware of those things which lead you forward, and those which hold you back."

-The Buddha

A mountain is defensive, while the valley is welcoming. At the top of a metaphoric mountain, as far as the ego could climb, the hiker's mind finds that it is still not quite as satisfied as earlier thoughts had promised. As he makes his way back down the trail, the disappointing thoughts build up a certain amount of rage that cannot be fully explained within him. So he decides to drop the hiking gear from his back momentarily. Grabbing a large stone, he throws it furiously over the side of a cliff into a large river. Unfortunately, as he throws the stone from over his head, he suddenly slips, and falls into the river which is rapidly flowing down the side of the mountain.

He quickly looks for something, anything to grab onto if he can, a rock or tree limb maybe. Anything in order to stop, and maybe get back to the top of the mountain, or at least get back to his gear. He is able to grab onto rocks on the sides of the river for short moments, but the river's flow is too strong, and he has to let go frequently. He thinks, "If I could just get out of these crazy rapids, I could get some rest." So he continues to cling to the side for longer and longer times, looking for a place to climb out of the river, while trying not to get swept too far down stream. Yet the flow of this river is just too strong, and the water has carved too deeply into the rock for him to be able to climb out. With his body heavily run down from trying to fight the flow of this river, he decides to keep clinging to the side, until he can think of a better plan.

Gripping onto a tree's exposed root, he looks towards the top of the mountain for a moment, then down at the valley. Suddenly there's a beauty to the valley that he didn't quite notice when he was there only to get hiking supplies for achieving the victory of climbing the mighty mountain. Suddenly the valley's delicate beauty is far more inviting to him than the rigid trek back up to the top. As this pauses his train of thoughts, he begins to observe the river itself, and notices it is deep enough to simply ride down towards the valley. "But will it be safe?", he wonders. "I better cling to this tree root a little longer and think about it. Will it get me back to the path I was on?"

Moments later, he becomes too tired to cling anymore, and decides to ride the river downstream, for now he thinks. After a couple moments of relaxing, and going with the river's flow, he sees the edge of a waterfall up ahead. Suddenly becoming more alert, he finds an opportunity to grab the edge of a rock, and climb to a nice flat surface with an extremely beautiful view of the valley below. With the wall on the side still too steep to climb, and the waterfall's drop not quite something he's ready for, he decides to stop and rest, for now.

There is no journey in life that is a straight line, no success without some kind of failure, and no going beyond ego without first being lost within it. It is very rare to become fully awakened right away. Just as the ego has built up a certain momentum within us, the ability to be present needs to build its own momentum. Very rarely someone like Eckhart Tolle for example, will have an instant awakening phenomenon. From the story of Siddhartha Gautama (historical Buddha), by the time he was fasting under a Bodhi Tree, he had already been on a long journey both internally and externally. Meaning he was most likely a partial blossom walking along the slow and steady path towards true freedom from identification with the mind. The Hindu culture he lived within helped to point him part of the way, but his own experiential journey was necessary for him to fully realize himself.

For most people this journey is like waking up a little bit at a time, and then falling asleep again a little less than before. A little more awakened as the present moment moves on, until enough presence awareness can increase beyond the mind's alluring momentum. There is no way of being more conscious than we are now. This is why someone who is not ready to see themselves beyond ego in any way, cannot understand what it means to be present on a deeper level than thinking about it.

You can suddenly find you're actually more present than you currently realized. This is likely what happened to Eckhart when his ego instantly dissolved fully. However, in most cases, until we reach a certain threshold of presence, moments of becoming aware that we are more present than we realized will more likely mean a significant increase of awareness that is still partially identified with thought. Either could happen at any time, slow and steady could turn into fully awakened at any time for anyone who is ready to fully realize, there is no time, only now. Fully aware that you are the present moment becoming more aware of itself.

While it is impossible to be more present than we are right now, the more we practice being conscious of our unconscious mental/emotional patterns when they arise within this moment. The more our presence transforms their lower vibrational frequencies into a higher level of consciousness that the mind's thinking cannot quite grasp. Everytime we notice inner resistance, we make it more recognizable, and we are dissolving it to some extent. Everytime we are aware of our inner spaciousness, we enhance our awareness of awareness itself, and become a little more awakened. A little more familiar with ourselves on this deeper level.

When the mind comes in again demanding your attention, or a challenge arises that the mind can't help but react to so quickly that awareness is absorbed too soon to be present. Simply notice, there's another pattern, or there's another challenge that life has presented. Learn to accept these challenges as if they are what you chose. Not saying yes to everything, but learning how to make decisions without the mind's negative patterns making them for you.

We don't become more present by trying to get rid of these patterns or challenges. We don't become more present by straining to be more present. We become more present by simply allowing ourselves to see things more calmly and clearly for what they are now. While noticing how the thoughts try to turn the present moment into a story of "me" that often involves a time based identity.

The river in the mountain metaphor is obviously the path into awakening, or simply the natural flow of life. From the rigid and defensive mindset of the mountain, that believes we must defend the illusory self image at all costs. This river is leading us towards the welcoming and forgiving awareness of unconditional love for ourselves and all beings in the world around us that can only be fully understood within the valley below, deep within us. The mind will try to cling to the sides of the river as much as it can, clinging to its conceptual self image. At the same time, convincing itself that clinging, holding onto the past, and resisting the flow of this river of life is what's necessary.

Whenever we notice the mind clinging to its identified habitual resistance patterns, we are able to let go from a deeper place within, and go with the flow down the path of least resistance. Until we are aware enough, the mind will continue to cling to the mountain's rocks whenever it can. The more aware we become of ourselves, and see how no matter what the mind reacts to, the river of life is always simply going with the flow. Down the path of least resistance, simply by being as it is. The more we realize that we are this river of life, we are life expressing itself as a human within this moment. The easier it becomes for life to guide us from a deeper place within, one that is not separated by forms. When we go with the flow, we often find ourselves on the path that is most beneficial at this time. Serenity towards what is doesn't mean there are no choices, it simply means our choices become more calm and centered.

Within the silent gaps between thoughts, an opening comes in. Allowing the opportunity for creativity and intuitive thoughts to come into this world. The more we experience these phenomena, and see the difference between an intuitive thought, and the repetitive thought patterns. The more easily we find ourselves on the most balanced paths in life. Not without challenges, life will always have challenges. The more we can welcome what is in the present moment, the more our human form becomes just the very surface expression of what we call life. The surface elements do not become meaningless, they actually become more meaningful the more present we are. From this deeper level of consciousness, life can guide us along our inner and outer journeys, allowing them to become one in the same.

There is nowhere to get to within the inner path, it is all within you now. There is no way of pushing the river rapids further down the stream than where they are now. Yet everytime we allow the present moment to be as it is, a deep part of us is fully awake, and we become the river. The present moment becomes infinite, life becomes infinite within this moment.

When the mind thinks peace is not here now, "because clearly my life is not as peaceful as it should be", or clearly the river's rapids are too strong to be at peace right now. Our attention is only temporarily pulled away from the deeper peace that is always within. The mind cannot let go of these patterns on its own, it requires our deeper awareness. The more we notice our own clinging to the side of the mountain, the more we are able to let go of the patterns. Our relationships with other people can help us see how conscious or unconscious we are. When people bring out unconscious patterns, they can help us become more aware of negative patterns.

After a nice rest with a beautiful view at the side of the waterfall, our hiker has reflected for long enough, and is now ready to jump over the cliff into the river rapids below. Much like the initial fall near the top of the mountain, he finds himself pushed downstream quickly, only this time he's ready to go with the flow. There are some obstacles as he rides the rapids, rocks to avoid, tree limbs that don't lead to a way out, but do get in the way.

Yet, unlike before, he finds himself alert and ready enough for what might come within this moment.

Gradually, the flow of the water becomes a little slower, and he eventually reaches a point where the walls of rock along the sides recede. Now he's able to climb out of the river to ponder what he should do next. His mind becomes conflicted as it thinks about the hiking gear that was lost near the top. "Is there any way I could get back up to that spot?" he wonders. So he decides to look around to see where he can go from here. However, with such dense, lively foliage, there are just too many trees around to figure out where one of the mountain trails might be.

Pondering the amount of time and energy it would take to go back up, and then back down. The mountain has finally become too defensive, while the valley is becoming increasingly welcoming. He decides the only beneficial path at this point is to continue following the river downhill.

Now walking into the valley, slower than the river's rapids were before, the view of the wilderness around him is able to truly sink in on a much deeper level than it did earlier in the journey. Slowly but steadily he becomes the valley, he becomes life itself consciously experiencing the deep connection with all forms of life surrounding him. Through the simple oneness of Being, he realizes this has been the truth throughout the journey, it was simply covered up by the obsessive drive of his overthinking. The need to have the mentally

274

defined "greatest journey" was obscuring the value of the journey he now finds himself on. Now there seems to be no reason to think about it. As he simply enjoys Being, while he continues walking his current path. He knows on an extremely deep level that life wants to find itself along this path. Climbing the mountain was a part of this, but no longer as meaningful as what is here now.

There is no path in life without your conscious awareness, there are no mental conditioning patterns unless there is a deeper consciousness to manifest from, there is no mountain without a valley to rest upon. Anytime you realize how rigid holding onto your conceptual mountain is, this awareness allows you to let go, and follow life's path of least resistance deeper into your inner valley within the inner body. To know the difference between the rigid and defensive inner mountain, and the gentle and welcoming nature of the inner valley, is to become aware of how the river (the natural flow of life) is not pushing you away from the top of the mountain. It is actually guiding you home, into your valley of oneness. Realizing that you are the valley of unconditional peace.

The more you bring acceptance into the life situations that surround you, the less you experience them as something that shouldn't be. The challenges that the river presents you with can be transformed from a conceptual curse, into a spiritual opportunity. An opportunity to practice serenity, and transcend the mountain like defensiveness of ego. Now every challenge in life has within it the potential for deeper awareness, whenever you can allow this one moment to be as it is. The river rapids will be rough at times, and smooth other times. Yet when you go with the flow, you become the river, you become the present moment, more conscious of itself than it ever has been before. Life transforms from an uphill battle into an alert presence that is ready to go with the flow.

EVOLVING EXERCISE

This might be the point in most books, if not sooner, when the exercise would be for you to write down a list of all the things you believe trigger reactive patterns, and keep you clinging to your conceptual mountain. While, you may find this practice helpful to some degree. Within a moment when resistance patterns have become active, it's not very practical to pull out a list of triggers, a list of how to stay calm, or even try to recite them from memory. When the mind has become more interested in reacting, maybe at certain times, it might be practical to wear something like a piece of jewelry or chant a mantra as a simple reminder to be present.

Yet, reactive patterns don't care much about reminders of reactions when their main motivation is to react negatively. It is only through experiential practicing of being present while these patterns react that can allow true change to occur. Then as a secondary aspect, reminders like jewelry, writing a list of triggers, or chanting mantras to ourselves, can become helpful for the more important exercise of simply watching. Allowing yourself to be here now, regardless of how much you might be clinging to a conceptual mountain of the current reactive impulses.

To simply notice your conscious state at this moment, is the exercise which transforms the way you experience challenges in life. From a point of view that is rigid and defensive like a mountain, to the infinite peace that is deep within your blossoming surrendered state, and welcoming to the challenges that the river of life brings.

A new level of awareness has come into your life that now allows you to notice your mental/emotional state from a deeper place, it becomes easier to allow these patterns to come and go. Utilizing a deeper intelligence than thought. Any moment you choose to ask yourself, "what state am I consciously in now?", you can now reflect in a way that the mind cannot do on its own. You have tapped into the source of creation, and the source of deeper intelligence. Simply asking yourself from time to time throughout the day, "how present am I now?", can allow you to bring more presents into the habitual repetitive patterns more easily as this practice continues. Simply by repetitively letting go of the mountain, you more naturally realize that you have always been the valley.

BEING
COMFORTABLE
WITH NOT
KNOWING

"The more in harmony you are with the flow of your own existence, the more magical life becomes."

-Adyashanti

Author of: The Impact of Awakening

The mind needs to know every detail. Is there a plan? What if something goes wrong, is there a plan for that? Will I find true love? Will I get married before I'm too old? What if we get divorced? Will I get this next promotion? What would happen if I got fired? What if I get sick and can't work? What if my car breaks down? What would happen if I stopped thinking about this? Who would I be if I stopped thinking about who I am?...

The future is always the perfect place for the unconscious mind to find problems because the mind can easily keep thinking endlessly and in a fearful way about the unknown. There is always a point when the mind is trying to let go of fear-based patterns that it will become resistant to letting go. Therapists find most people who believe they want to change a behavioral pattern usually come to a place within their therapy where they are defending the patterns they say they want to change. Simply because the ego fears the unknown almost as much as death. If anything is not within its illusory control, there is a defined problem. Meaning the future can always be food for fearful thoughts, and more thoughts, and more thoughts.

It's not so much the future itself that is feared, but the lack of self image. What will happen to the conceptual "me"? Deeply programmed into our habitual thinking patterns is the ego's need for survival through holding onto our habitual thinking patterns. We may agree that we want to change a certain behavior, or even change the way we look at ourselves and the world around us through dissolving ego. The only catch is the momentum of habitual thoughts that think they want to change, until real change comes. When fear of losing the familiar outweighs the urge to find something new, the awakening process can mean losing ourselves in this fear again and again, until deeper awareness outweighs these familiar patterns of fear.

Social media has many uses and potential uses. One is simply entertainment, another is showing pictures of ourselves and family. Today one of its main uses is memes and posts that display just how important it is supposed to be to feel anxious, fearful, and angry. How important it's supposed to be to feel right by proving others wrong. Very few are using it as a tool to learn something new, or go beyond a self identity.

Most people go on social media to strengthen their ego while telling themselves how important it is to somehow find victory over the opposing viewpoints, which are simply made up of other humans who are telling themselves a similar narrative. "I'm not going to back down, others will have to see how wrong they are!" Which most likely strengthens the other person's opposing view, rather than helping them see something new. Imagine what social media might be like if people log on in order to learn something they didn't already know. To search for new viewpoints on a subject, instead of reinforcing what they believe to be true or identify with in the first place.

Is it even possible for that time to come? When it does, social media will become the town hall that it had originally advertised itself to be. There will likely be more in-depth conversations with people currently on any side of a conceptual outlook. As opposed to arguing arbitrary ideas and name-calling for the sake of strengthening the current identity with an opinion. A place to learn new details as opposed to a place to strengthen one's argument. An argument that was most likely gathered from other people's thoughts, who also gathered their argumentative points from others. As 99% of our thoughts come from outside sources, then become identified with.

When we are able to live more in stillness, deeper understandings naturally come in between the gaps of thinking. It is impossible to hate another human being when we experience empathy. It is impossible not to experience empathy when we are not lost in overthinking. Like anything else, social media is not good or bad but a reflection of how we use it. A reflection of our current level of consciousness. In the future, social media might become more of a source of knowledge and wisdom as opposed to a source for strengthening egoic anxiety. I imagine there will still be plenty of images of cats doing crazy things as well.

There is no mental agreement that can point us beyond identification with mental agreements. We cannot solve a problem using the same level of consciousness that continues to create it. We can have various amounts of success in the self help department. Starting new mental habits like always thinking positive. It can be great advice to practice thinking positively instead of negatively, unless the practice itself becomes too insistent that only positive thoughts should exist. This not only pins us in an impossible situation, it demands through inner resistance that positive thinking is a problem that needs solving. Feeding the inner resistance we are attempting to overcome, and therefore will guarantee we become negative while trying to practice being positive. The universe is too paradoxical to not invite the full spectrum into the practice.

While positive thinking is helpful, it is just as necessary to allow negative thinking to exist in order to notice it without suppressing anything, or clinging to a self image that insists, "I don't do that sort of thing anymore", while continuously doing that sort of thing. Not all, but most self help platforms help enhance the egoic self. The mind might create some new behaviors, at least for a short time. Yet more so, they promote a new ego self image. Exercising the mind's narrative of how, "I'm gonna do it", and in many cases, at a later point, they can also refresh thoughts of, "I knew I couldn't do it". Usually without the ability to see how a new self image will at some point, bring back many, if not all of the old patterns. Because ego created them unconsciously in the first place, long before we changed little details of this imagined self, the gravitational pull so to speak, will very likely find its way back towards the longer practiced patterns of the time based "me". Only by dissolving time/ego through awareness of the illusory identity, do we move beyond the dysfunctional patterns that this creates. When we understand this within ourselves, self help platforms might bring in additional support. You may find one, or some that are very helpful. Yet simply allowing ourselves to be as we are now is the most helpful thing we can do to help any living or life situation.

Most self help narratives are projecting a future self that will be an improved version of a mental identity. Not to say we should throw away any self help book we have, or never learn any new skills because it might enhance the ego. Only that there can never be a future version of you. When it comes it will be the present. Future goals are great, and helpful for so many things in life. Spiritual awareness is the one area where it hinders our progress. Because our progress of becoming more present always relies on this moment now. The most truthful essence of who you are is unchangeable, and infinite. There are many things we can change on the surface of life, but nothing needs to be changed in order to know yourself more deeply, and therefore bring joy into every surface situation in life.

In this one area, accepting what is here now, is what creates a better future. A future of being more at peace within this moment, which will only come as the now. This self help has a brighter future while paradoxically not having a future. By simply using whatever comes into the present moment as a practice of simply noticing when the ego presents itself. Life, or the present moment becomes increasingly liberated from ego patterns. Self help books can be great, consciously noticing the ego trying to reestablish itself within these new practices can be priceless.

There can be a future goal without clinging to it, without clinging to the self image within the thoughts of a potential future. When we practice being comfortable with not knowing, the potentialities become stronger. We can basically get out of our own way. We can choose to do something, with the added ability of consciously choosing not to worry about the outcome. Maybe it will happen as we wanted, maybe not. Maybe something better comes along, either way things will work out. This can be very helpful for goals set in the future, but like anything else that goes beyond ego, it requires practice.

To become comfortable with not knowing the labeled big things in life. First we need to practice becoming comfortable with not knowing the labeled little things. Will I get to work on time? Will there be a lot of traffic? Will I get a good parking space? Will they have what I'm looking for at the grocery store, or will they be out of it like half the time I go there? Will it rain later today like the forecast claims? Will the wash get this stain out of my shirt? Can I fall asleep tonight, and if so, will I wake up and have insomnia again?

Worrying pretends to be useful, but is always projecting, or asking for the thing it is avoiding. Always looking for chaos in some way. We cannot control whether or not there's a good parking space for example. When there isn't, the mind might say, "Of course!", reestablishing that it's the only way it could have happened. When the space we wanted is available, the mind might be surprised, as if it must be a once in a lifetime flook. Unconsciously using the negative thoughts as a way of clinging to the identification with the inner patterns of resistance. The ego doesn't care if its identity is mostly a negative one, just as long as there is a mental identity.

An incredible amount of freedom comes with the ability to acceptingly say, "I don't know what will happen". Allowing an opening for something deeper than thought to come in. Resistance based thoughts are mostly looking for more resistance. To surrender to this fact, not try to fix it, but simply allow it to be as it is, is to allow serenity to transform resistance into acceptance. Probably not all at once, but this is the beginning of the end for inner resistance. No inner resistance can survive your full acceptance of it. Accepting inner resistance exactly as it is now. Is accepting that we don't quite know all the details of how it dissolves from this point, or how long it takes for our unique awakening path to fully shift beyond ego. However, the more we accept where we are right now, the deeper we will find peace within the future. Not by getting to some imagined place in time, but allowing the future to come to us, the present moment. Any conceptual idea of the future becomes secondary when we can fully accept this moment as it is now. Working entirely on this one step we are on.

Insomnia can be a perfect example of a problem the mind tries to avoid while creating it at the same time. Odds are, if the mind is worried about not being able to sleep, it will find itself awake in bed thinking about why it can't fall asleep, and all the other problems it's certain it needs to think about while lying awake in bed. The one place where there is zero need to think about "my life", the mind cannot seem to stop thinking about the time based version of "my life". "Why did that happen? Why won't I do that? Why can't I do this? When will my life start working out for me? Why can't I stop thinking, and just fall asleep?"

The more we resist resistance, the more we reestablish resistance patterns. There are two types of insomnia, one that is resisting time, and one that is accepting the present moment as it is. The thought, "when can I just fall asleep already!?", will insure it is very difficult to do so. Because it excites and invites the negative thoughts and emotions to become more and more excited. However, the conscious acknowledgement of the mind asking itself "when can I just fall asleep already!?", adds a deeper layer of awareness that changes the situation. Insomnia can become a perfect moment for meditation if we can choose to let go of time, and practice becoming comfortable with not knowing. By bringing awareness into our breathing, a space opens up that allows us to step back from the thinking mind, and watch it from a deeper place. Suddenly we are no longer fully trapped in thinking, a calming spaciousness is helping the mind and body become more relaxed, bringing in more potential to fall asleep.

I have experienced both versions of this situation, and have found that even if I still can't fall asleep, accepting the situation as it is allows me to feel more rested when it's time to get up. Simply by not having to be in a panic mode from a time based problem. Not demanding any specific outcome, but allowing what already is within this moment, to be as it is. Just by bringing in awareness of breathing, bringing in awareness of inner spaciousness, creates a much more peaceful scenario. The mind may grab attention for a moment, then we remember to focus on our breathing, and this can go back and forth like many meditations do. Yet any amount of presence will make any situation more peaceful.

I've found that the more we can inhabit the space that is outside of thoughts, the more likely we begin to notice the mind falling asleep. The feeling of the brain producing its natural chemicals that slow down thinking may even become recognized as we fall asleep. This can only happen when we become comfortable with not knowing, and simply allow ourselves to be here now. Maybe we fall back asleep, maybe we don't. Either way, practicing accepting the situation as it is means there will be less stress than when we are resisting what already is.

In any situation throughout the day, the mind can become uncomfortable with what it cannot see, or what is beyond the grasp of knowing at this time. This is a strong evolutionary accomplishment that enables our species to become as successful on the level of mind as we have, and helps all the creative forces we experience. Except, much like a drug, what is helpful within smaller doses, becomes more problematic with increased use. It's one thing if needing to know brings about a new idea, or new technology into this world that is helpful. It is a whole other thing if needing to know only creates an endless stream of suffering based thoughts that are certain the worst case scenario is sure to occur. Yet paradoxically, the resistance based need to know can be the greatest agent in moving us beyond resisting what we don't know. What we resist will ultimately lead us towards the acceptance of not knowing. Not for everyone in their current life, yet certain individuals, and therefore humanity as a whole, cannot help but reach this conclusion when we become ready. For our species as a whole, inner resistance is our greatest teacher in finding the next step in our evolution.

The certainty of negative results can find its way into any and all life situations, until you can take away the element of time. The more you practice being comfortable with not knowing, the more you realize how little we actually need to know within the simplicities of this moment. Education is important, but knowing everything about everything is extremely delusional.

The mind may think that all the secrets of the universe will be achieved when we become Enlightened. Really what we achieve is the lack of needing to know everything. Needing to always achieve a very specific goal. A goal can be an extremely beautiful thing to have, clinging to a specific outcome is in itself a form of insanity. If awakening achieves anything, it is the dissolving of conceptual time, and a renewed wonder for what the future might bring to the present moment. The future is no longer good or bad, it is simply the present moment that is waiting to occur.

Evolving Exercise;

Life is always one step at a time, including the so-called big events that the mind becomes worked up about. Every element has to be just right or the entire event will be ruined. It is extremely liberating to plan for what the mind labels as a big event, without being lost in the future along the way. To be in the present moment while figuring out all the details like what to wear, who to go with, or how to get there. To not be taken prisoner by the mind when it finds something has gone wrong along the way. As some element almost always seems to not go as planned.

The journey itself becomes transformed from the mind's unconscious need for a future problem. Into a joyful wonder of how the entire journey plays out one step at a time. When there is no past or future, no before or after the big event. A deeper joy can be found in the wonder of what the future may bring to the present moment. As less resistance is encountered, more presence is placed on every step of the way. The entire journey becomes more meaningful. Major problems are less likely to occur when our own inner energy is no longer constantly expressing resistance towards the unknown. Unconsciously looking to reflect this energy off of every situation we encounter, more easily attracting what we try to avoid, identifying ourselves with any problem the mind defines.

The more you inhabit this one moment with a deeper presence, the more the journey is allowed to be as it is. Allowing it to be perfect regardless of physical outcomes. Practice this in your daily routine, and the so-called big events in life can become just as simple. Any moment that no longer needs to be attached to a conceptual future outcome, is an extremely peaceful, and joyous moment.

LETTING GO OF PAST

"Thinking is just a recycling of data that you gathered in the past."

-Sadhguru

Author of Inner Engaging: A Yogi's Guide To Joy

While the future is ungraspable, the story of the past is filled with memories that can be extremely difficult for the mind to let go of. The ego's story of "my life" is all that an unconscious mind can understand itself to be. Because without the story of time, there can be no ego. What we could call the historical story of our life can include time without the invested mental identity within it. All the things that we've experienced in the past without a need to cling to or defend a mental self image around. The ego likes to reestablish itself by reacting to past events as if they are here now, because like any expression of life, the ego wants to survive.

When we notice the mind's clinging to past stories, we have gone beyond ego, which cannot survive your awareness of it. It can be easier to see these patterns in others first, yet until we can see the ego within ourselves, the ego of others can keep sustaining our own. What we haven't yet made conscious within ourselves will continue to repeat until we do. Other people do not create our patterns, but help trigger them with their own unconscious reactivity.

All that is needed is the agreement that there is no past within this moment, there is no part of you that is in the past at this moment. Awareness of past patterns may take time, time being present. Yet the understanding that there is no time, is what cuts away the cancerous tumor of illusion that humanities ego has become. To understand deep within that any story of the past is not here now. Is the end of being fully lost in the mind's obsession with time. The need to be there instead of here. The more we simply notice the mind clinging to the past, the more we move beyond the clinging. The light of your presence dissolves the darkness.

Conceptual time plays a valuable role in our lives, and benefits us in many ways. Unless or until the obsession with the time based "me" turns a story into a crutch as the mind tries to cling to, or recreate the past. "I was this", "I was at this place", "I knew that person", "I was never what they said I was". In many cases, a story of the past can quickly become toxic fuel for ego. Fuel for investing interest and attention into a mental story, away from the here and now. Needing an attachment to something, anything that might take us back just even the slightest to the experience of that earlier present moment. Ironically pulling us away from our true self, our deeper self. Simply bringing presence into this clinging, is what lets go of unconscious clinging to the past.

Of course there are also the past events we wish we could stop remembering, but the more we resist them, the more we seem to remember. Because we can't help but reestablish an identity through our resistance towards the memories every time they present themselves. The more we can surrender to what is, and bring acceptance into this situation, the less likely a memory of the past can feed off resistance patterns now.

Whether a memory is considered good or bad, it is the amount of an invested self image that keeps it by our side so to speak for as long as the memories grab our undivided attention. Noticing when the mind is trying to get, or is in the middle of getting this ego based attention is the beginning of the end for a time based mental identity. Thoughts will continue to present themselves, and may continue to hold a certain weight for some amount of time. However, the less invested we are within them, the less egoic, and less frequent they can be. This allows memories to become more and more useful, while becoming less and less dysfunctional. More influential for learning and maneuvering our life in conscious ways, and less of a weight that holds us in dysfunctional patterns. Holds us in a past that no longer exists, and can no longer be who we are.

A great example of historical time within my life would be the time I spent in the U.S.Navy, living on board the aircraft carrier USS Abraham Lincoln. I worked in the hangar bay as an Aviation Boatswain's Mate Handler, where our main tasks were moving aircraft around in the hangar bay, as well as on and off of the elevators that moved to and from the flight deck. We were also the main fire control if any fires were to break out in the hanger bay. There were plenty of pros and cons that the mind could analyze about this living situation.

On one hand, it was fun moving jets around, often inches from each other, in a large hanger, on a surface that continuously lists slowly back and forth, port to starboard, as the ship moves around in the ocean. Living in the middle of the ocean for a large percentage of the time, on a giant steel ship with a population of roughly 5 to 6 thousand people. On the other hand, living in the middle of the ocean for months at a time can become stressful in various ways, especially knowing that we were involved in military combat operations.

Today, the military has many reasons for why it needs to exist, some are more truthful than others. Some are simply a result of still living within humanity's ego stage, still living in the ways of our past. Still needing to protect ourselves from ourselves so to speak, as the unconscious superorganism that we collectively are, until we can collectively awaken from this state. Becoming more ready to see that at this point in our global society, only through abundance of resources and freedoms for every human being, will we truly find the most successful survival strategy for humanity. Battling each other is the least effective way of achieving this. Yet the ego often needs a battle against something, anything. It doesn't matter much what our inner battle is with, as long as the inner tension is strengthened, the ego feels complete, more real in some conceptual way. For at least a brief moment, until it needs something else to battle with. As long as we cannot see this process within ourselves, it will always be mentally blamed on an external situation, person, or people.

This is why some sort of grudge, or feud is often needed for a national ego to strengthen itself, creating chaos that isn't necessary. It's necessary to the ego, but also helps the endless patterns of warfare to continuously re-perpetuate themselves. For the ego, the military is a practical solution for protecting the identity of nationality. Perhaps without this need, there would be far less national defense, and far more cooperation attempts with other conceptually declared nations. At a time when most humans identify themselves through concepts, a military is still a practical thing to have. One day, in a world that has dissolved collective patterns of clinging to a conceptual past. There will be no use for national defense, there may still be nations, definitely still cultures, but with less of a need to define ourselves and our world through conceptual labels. It becomes natural to work together when there's no more need to define and resist an "other". When we are content within ourselves, living in peace is the most natural way of being.

Growing up, my mental identity was not anti military, but was also never pro war. So I never thought I would end up in the military. This outlook may have helped me observe the military way of life without fully identifying with it. Not that I would have been able to understand this at the time. When I dropped out of high school and decided to get my GED, I was encouraged to join in order to get college funds. This was my main motivation, and hoped to see four peaceful years. However, I was stationed on the carrier in Everett, Washington, for about 2 or 3 months when the terrorist attacks of 9/11/2001 took place. This likely added to the normal stresses of living on a Navy ship. Yet life on a ship, regardless of training or real combat, is seemingly the same business as usual. Nowhere near as stressful, or life threatening as what ground troops must have experienced.

From my perspective, there was always a sense of empathy for all humans on all declared sides of this conflict. I could not quite relate with the grudge people held against the Terrorist groups, especially as I could see how this led to some level of insanity for those holding the grudge. I understood the need to react, I understood that terrorism should be resolved in some way, but not the need to hold a strong hatred over people I didn't know anything about. This hatred was especially heightened within those who entered the military after 2001.

I can't say that I noticed this because I was more conscious, because there was still plenty of confusion within myself, about myself. Yet on that subject, my ego did not lower its level of consciousness into hatred towards this defined "other". Perhaps at the time, my ego simply had plenty of inner stress through self judgements, that I simply didn't need this identity as well. Perhaps I'd learned enough about past military conflicts throughout history to not find myself lost in this type of grudge. There is really no way of fully knowing, as it was the past, and I am in many ways no longer the same person.

I viewed terrorism as a form of insanity, and was unsure if it was being dealt with correctly through our military responses. Yet I also understood that I agreed to do the job, so I simply did my job regardless of the paradoxical psychological complexities that were creating so many great questions within myself. My stress was clearly nothing in comparison to someone in other situations, like face to face combat. Yet being at war in general will hold a certain level of stress regardless of one's tasks.

At the end of a six month cruise that was extended to a nine month cruise as the Iraq war was starting. The Aircraft Carrier I was on got a visit from President Goerge W. Bush as he was flown onto our ship just before we dropped off the aircraft squadrons in San Diego. I was also on the flight deck when he gave his speech declaring major combat operations in Iraq had ended. At least they should have ended, achording to his thoughts. Unfortunately the insanity of this war had only just begun. However, my time in the Navy ended before I would have been in the Persian Gulf again.

This is a truthful, historic story from the past that is no longer seen as a part of who I am. It is relevant within my historic journey of time, but no longer relevant in identification with the story of time. I no longer need to imprint a story of "me" into every story of time the mind thinks about. Allowing the stories to no longer become more about myself than they need to.

Now stories of my historical past can reestablish the acknowledged progression of awakening out of conceptual time, instead of reestablishing an identity with the past. I am also no longer gripped by a political opinion of president Bush as I simply explain the situation involving him. I could say that I still hold political views, but they no longer hold the heaviness of an identity around the views. Our politicians are the most popular egos as much, if not more so than they are the most appropriate person for the job. This will not change until we collectively do. Finding acceptance with this can be very helpful in not strengthening suffering through political opinions.

To simply notice how the mind holds a certain nervous like, or anxious like energy while thinking, or talking about the memories of "my life". To simply notice how the mind wants to desperately cling to the familiarity of a memory, is to be free from the mind's clinging to the good old days, or the nightmares of yesterday.

You cannot be there when you're fully here now, feeling the peace within the inner body now. This can take practice, but there is no past within your presence here now. The story of time is merely a story, when any amount of presence can acknowledge this, there is less and less need to invest a sense of self in something that does have a certain historical value, but is no longer needed to be a part of who we are today. The historical story of you can have many meaningful and practical purposes. The only purpose for the ego's story of "me", is to sustain its illusory sense of self. Sustaining the suffering it is familiar with by repeating the same patterns. Repeating the same thoughts and emotions when bringing up these stories of the past.

The forms and situations in life are always changing, and always will be. Yet it is your presence that time can never alter, it is your presence that brings light into this world. Presence awareness builds with practice, but it is the same level of deep awareness at every moment along the journey. Simply notice when the mind is investing a sense of self within a memory, and truth shines its light onto the illusion. Simply continue noticing these patterns when the stories come into the present moment, and the patterns will lose your invested identity. An egoic investment in the past can no longer be fed if we are no longer convinced that the past is where we need to find ourselves. Allowing the timeless you to become more recognizable, and allowing life in general to become more peacefully balanced as the stories of the past no longer control the now.

The deeper "I" comes into play from this point on. Anytime that life requires a story of your past, a new relationship with the stories has been born. Whether it's going into depth with a friend about your past, or simply mentioning your work history at a job interview. The stories are the same as they would be. Just no longer tainted with an added self image that will try to derive today's sense of self from that earlier sense of self. A new freedom is born where the past can be remembered without any dysfunctional side effects that the egoic mind needs to unconsciously produce.

No more need to display a highly micromanaged version of who you were, and who you are within the story of time. The mind may still think it needs to present a certain image to others, yet without an invested identity, thoughts are reduced from commands to requests. It becomes easier, and more natural to laugh at ourselves in healthy ways, and less likely to be embarrassed in unhealthy ways. There is nothing right or wrong about embarrassment, to resist any emotion is where dysfunction comes in. The ability to laugh at ourselves is true freedom of self judgment.

When the story of the past is no longer identified with, our relationship with thought is renewed. Thoughts become more like the sidekick of our journey instead of being the main character. Not every thought is useful, and many can be problematic. As the main character, this creates a lot of problems out of simple situations. Yet when our stream of thinking is placed in the sidekick position, thoughts can become more useful, because our deeper awareness of the mind is what it needs in order to become more balanced. Allowing thoughts to take on a much more fitting role. As Being becomes more relevant than thinking, choosing which thoughts to follow becomes more relevant.

Attachment to the past keeps ego alive. The complexities of the layers and layers of psychology make the stories difficult to put down. Any thought about our past can trigger many more. Any perceived issue we feel we still have with the past can create complexities within the thinking in order to keep thinking, and keep identifying with or against the narration of the mind's story of time, story of "me". Once we see this, it becomes very simple, there is no way to think our way out of problems created by thoughts.

Simply seeing how your thoughts can cling to the past in a problematic way allows the complexities of overthinking to transform into the simplicity of being with what is and accepting what already is. The simplicity of watching thoughts, simply being with the thoughts through presence. Your awareness shifts from being past based, to being rooted in the now. Shifting from being lost within resistance to becoming more aware of your deeper acceptance. The mind may disagree with what it can't understand, but that's okay too, when not every thought needs to be followed. Simply acknowledge that there is no past within this moment, and no illusion of time will be able to cover up this truth. The present moment is all that life ever consists of. When this becomes clear on a deeper level than the mind, you will no longer need to look for yourself in a place that no longer exists, and your true self, the infinite light of presence, can shine through the cloudiness of overthinking. The historical past still holds value, but who you are now is infinitely greater than a distorted image of who you once were.

Evolving Exercise

Deeply knowing that there is no past within this moment, every time a memory comes, a deeper level of consciousness is now involved. Nothing needs to be done except for watching. Thoughts cannot watch themselves in this way. The moment you can observe the mind's thinking, you are bringing a deeper level of consciousness into this world. Empathy for the past now has space to emerge in between thoughts. There is no true time based nightmare when we introduce this deeper acceptance into the aquation. Practice feeling the presence within the inner body, become anchored in the now. Anytime you can be present while thinking about the past, the past is letting go on its own. Nothing will be lost, you will still be the same you who experienced these past situations. You simply no longer have a need to lose yourself in a place you cannot be found, allowing you to find yourself more fully within the only moment you can.

Ever Evolving Humanity

"The universal order and the personal order are nothing but different expressions and manifestations of a common underlying principle."

-Marcus Aurelius;

Meditations by Marcus Aurelius

The earliest hominins faced the challenges of becoming more intellectually conscious, through a need to out clever large predators who were hunting them. Creating a means to become hunter/gatherers themselves. As well as creating a need to cleverly compete with other hominin tribes and species. Today we face the opposite challenge, we have become too clever for our own good. We have become too absorbed by the dysfunctional aspects of the overthinking mind for our own good. Now if we want to survive as a species, we need wisdom to outweigh cleverness. We must become more aware of the conceptual suffering that unconscious thinking cannot help but create. This simple situation continues to bring billions of humans closer to the point of readiness for becoming more conscious than we currently are collectively as a species.

As the seemingly lush and pleasant forests of endless conceptual thinking continue to transform into an environment of endless self-created mental and emotional suffering, both individually and collectively. Each generation becomes a little more collectively ready to find the next step in our evolution. Creating the current need for consciousness on this planet to grow to a higher level of awareness than the realm of identification with thoughts and emotions. To a higher level than mental cleverness where consciousness becomes aware of consciousness itself. Consequently becoming aware of the infinite peace beyond thought that is deep within us all.

At a later present moment, humanity may look back, and be extremely grateful for all the seemingly endless eons of suffering that humanity created for itself, which came to an end when it was able to allow this transformation of consciousness to occur. Understanding how it was this process of suffering which allowed all of humanity to find true inner peace. Once suffering has made us conscious enough of ourselves to no longer be able to create unconscious negativity, it has served its purpose, and humanity can move forward in peace. Then, Earth becomes a more conscious planet through us. Wisdom will outweigh the need for cleverness, and compassion can finally outweigh the need for selfishness.

The problems of the world will seem to almost fix themselves when humanity is ready and willing to do the work, without confusing ourselves about what is needed. Will it even be considered work? When we are fully present, meaning and purpose are easily found within the simple actions we take in the now. Through ego, every step towards progress or change, is resisted as much as it possibly can be, by anyone who holds an opposing mental viewpoint.

All our efforts are placed in resisting this moment as it is. Even the ego's approach towards positive change must hold some form of resistance within it in order for positive change to make sense to the ego level of consciousness. It must be expressed as a fight against something or someone to some degree. In a world beyond ego, change can be celebrated in more positive ways without the shadow of ego lurking around the participants. Social and economic progression that is universally beneficial for all will become much easier to create. Humanity can finally look for the paths of least resistance, and allow the world to balance itself. Without the dysfunctional unbalancing of the process at every step along the way.

Nothing in the future can be predetermined. However, where we are now collectively is not very likely where we will be just a few generations from now, if not sooner. The near future is likely to be a little rough, as real change usually is. Depending on how much drama our individual/collective egos need in order for us to become aware of what ego actually is. Yet this too shall pass. We cannot stay locked in these levels of suffering

307

for too long before something new starts coming into the equation. After all, the larger reason why our world is so seemingly insane today, is out of our need to move beyond the mind's insanity it cannot help but create for itself. Somewhere deep enough within us, all of humanity knows this truth. We are just simply at various levels of unconsciousness. One thing will always lead to another, change is always inevitable. If the dream world is only a little unpleasant, there would still be little to no reason for us to move on to the next level of consciousness. The more we create suffering, the more we create the readiness for shifting beyond it.

All of our ancestors who experienced egoic suffering have allowed us to get to this point in our evolution. Their suffering may seem like it was pointless, except that you are the pearl that is the next step of conscious awareness on this planet in which their suffering has helped humanity create. Your acceptance to what is, is in fact the very thing our planet needs at this moment. Your individual serenity, along with other awakening humans, accelerates the collective inner serenity process. The end of human suffering on Earth literally begins with the end of your suffering now. You and many other individuals are the universe becoming more aware of itself.

There are two movements happening within humanity right now. The movement of ego, and the movement of awakening. If we watch the news tonight, we will likely only see the ego movement. Mostly because that is still the most prevalent, but also because the ego is a lot louder than the simple silence of Being. Those awakening make less noise about themselves, and the world around them. So far, when someone within or beyond the awakening process tries to point towards the simplicity of inner peace, it is extremely difficult for most people to notice this peace they are pointing towards when the normal mindset needs to prove just how chaotic the world must be in order to continue strengthening the suffering they identify as who they are. However, at some point the awakening process will become noticeable in the news, and in what we call pop culture. More and more elements within society will begin to reflect this new movement of humanities collective awakening.

The media, politicians, owners of giant corporations, and any other establishment including religious organizations. All require the attention and cooperation of the general public in order to continue on along as they do, with business as usual. Which one of these might need to shift their way of life first is hard to say. However, eventually even religious structures that can seem to be the most egoic in many ways will have to adapt in order to continue on as an established organization. Luckily, most modern religions are not egoic in their roots, but simply egoic today because of where humanity is at this time.

Looking at the history of Christianity for example, you can see the build up of ego alongside the misunderstandings of Jesus' teachings shortly after the first proclaimed Christians. As deeper understanding turned into another form of ego identity. This understanding, and misunderstanding doesn't occur in a straight line, but has shifted back and forth many times depending on the level of readiness to look within for those involved. As it was mentioned previously, the beauty of this is that no one who identifies themselves as Christian in an egoic sense has to simply throw away all their beliefs once they begin to look beyond mental identification. It also means we don't have to stay within any religion if it no longer speaks to our current path.

Some traditions will change, some will become altered, some may end, and some may become strengthened depending on how well any individual custom can help point us towards true salvation within ourselves living within this moment, as opposed to placing a large emphasis on being favored by an external entity that only allows us to have a good afterlife if we live correctly according to our understanding of the rules we are given. These rules transform into guides or suggestions that can help us find the sacredness that is within us now.

With this deeper understanding, the true intentions within the teachings of Jesus (as well as within other religions) becomes more clear, and in many cases can strengthen one's faith on a deeper level, instead of looking at Jesus on the cross for example, as a means to create suffering for ourselves because of the ogo's interpretations of "original sin". Feeding the heaviness of ego through self judgment, and keeping a misunderstood hierarchy in a position of egoic surface power structures in the ways that ego understands and creates for itself. Original sin can become more understood as simply being born with the seed of ego, or the root of suffering within our genes. No one's fault in the ways that the psychology of ego needs to interpret it, but simply one of the challenges of being a human on Earth. Original sin can simply point towards where we have been as a species, and the challenge of awakening beyond this level of awareness. Not ruling out any afterlife elements, but focusing more clearly on the only place we can be now. Death is no longer the focal point when living more deeply in the now is the stronger message.

All religious establishments are at various degrees of ego. Most religions tend to pivot back and forth from being strict and conservative, to being more open minded and liberal. From being more form based, to being more formless based in their teachings. Depending on the current culture that the teachings reside within. Many modern Muslim communities in the Middle East could be viewed as a completely different religion than what it looked like during what is known as the Golden Age of Islam. Between the 8th and 14th century, the Islamic empire was leading the world in their understanding of science, as well as many other cultural and societal aspects. Beginning its decline with the invasion of the Mongols in the mid 13th century.

For around five hundred years, Baghdad was one of the world's largest cities. With a cultural mindset that was so open towards new ideas and understandings, that they searched far and wide for any knowledge and wisdom that they may have not yet known. The complete opposite of what muslim culture in the Middle East looks like today, which appears to be sure to have all the answers from the current interpretations of the Quran. This is at least true from the point of view of the current power structures who only know how to rule in a heavily egoic way.

The Golden Age of Islam is credited for creating many things like the first magnifying glass, the first fountain pen, as well as the earliest forms of chemistry as far as we know. Proving that it's not so much the religious doctrine that molds a society, as it is how the doctrine is viewed and used within the society.

In the future of humanity, a church, synagogue, mosque, temple, or any facility labeled as a place of worship/spirituality will only continue operating as it does, if it can continue providing a service for the awakening public. They may very likely come to a point where their services are mostly focused on the spiritual development of children, teenagers, and young adults. Attending services until they reach a point of deep enough self understanding that they can move along their own path, making spiritual growth decisions for themselves. Maybe coming back from time to time for reminders, and further guidance.

Yet no longer bound to egoic power structures of humans who believe this to be the only way. Perhaps these facilities will use many different ancient and modern methods that can point for various people at various different positions within the awakening process, as opposed to the one size fits all method of today's religious structures. These types of spiritual centers would understand the importance of letting go when the work is done. They will also not be seen as mandatory, as many people may choose not to attend any of their services. Many people may choose to work on spiritual growth with family members, friends, or simply by themselves.

These facilities may be mostly focused on youth, but will also likely be open for spiritual growth and wisdom for people at any age. Families and communities may congregate from time to time for certain celebrations, no longer as a reinforcement of ideology, but as a reminder of how helpful these facilities can be as well as a way for people to simply come together in social celebrations, which is an element within many religious gatherings today. Within this future possibility, celebrations of a spiritual community would no longer revolve around a lifelong devotion to any specific intellectual paradigm. It is hard to imagine such an organization as it would look like something out of this world, but would become the simplest of conclusions in a world that is truly moving beyond our current collective level of consciousness.

Suppose this is what a religious service might provide within an awakening humanity. Those who are finding inner peace will also live in a world that collectively promotes awakening from ego in any and every other way that ego strengthening is promoted today. Today the news affiliates would lose a lot of viewers if all they reported were positive stories. Usually, they only spend time on a positive note in order to take a short break from all the chaos and tragedy that the public is addicted to learning about. This is not saying they shouldn't report the madness of today's world or that they should report everything with positive rose-colored glasses. The news should report the true events that occur. What is usually most distorted in today's media is how often they can not simply report what is in order to keep ratings up. Most of the time they try to hold a certain narrative, and especially with a certain amount of negative tone as they report the current events. Even within a story that has a positive light on what has occurred, there is almost always at least a hint of suffering involved. The positive note will get buried by the standard tone of negativity, as most stories are designed to keep the viewers' negative feelings glued to their content.

This, like any other ego aspect of society, is an unconscious mechanism, and requires our conscious awareness of it in order to change the patterns. In a world of growing consciousness, the media will have to shift its performance in order to continue to reach a shifting audience. This may be confusing for some time to those who are not yet awakening. Yet at some point the news will no longer feel the need to paint a story with negativity, or positivity. The need for truth with wisdom within the information will outweigh the need for dramatic effect, or even propaganda. The censorship of what gets reported will likely change in certain ways as well.

Today a news story must go through the proper authorities before it makes it on the air, or published as an article. If it doesn't conform with a certain paradigm structure, its format is changed until it does, or it gets rejected. This can be the network's opinion for what is seen, its sponsors' may have a say, or it can even be the federal government's policies that decide. This can make it easy to start looking into conspiracy theories relating to the sponsors', as well as the government's affiliation with the media, and some of these can be true. More importantly we can simply step back and notice the mechanisms of ego at play, and become aware of how thoughts about the collective ego might be pulling us back into our own patterns of drama and chaos.

There is no moving beyond ego by worrying about ego. When we can simply notice, and accept it for what it is, the cultural/collective ego expressions of society can no longer pull us into thoughts that strengthen our own. At first you may want to stay away, or limit your exposure to negative based media information. Yet at some point, you will be able to see the elements of ego within the media, without being pulled into your own

315

patterns of resisting what is. Misinformation is not as big of a deal when the mind's overcomplexities become as simple as knowing ego for what it is when you see it.

We can still be misinformed, but we become far less likely to create suffering through any particular narrative, and far more likely to use critical thinking with the information we're given when we no longer attach an identity to it. This is how society will steadily become less vulnerable to being misinformed. We will no longer need to attach ourselves to a narrative in unconscious negative ways, and be less likely to identify with propaganda.

When we identify with a narrative, critical thinking skills are the first thing to be lost. When we simply see ego for what it is, we no longer need to fight for or against any opinion, allowing new thoughts to come in, allowing new questions to be asked about the current situation. A need to win an argument, or to hold onto a problem, turns into a joy for curiosity. Truth becomes more important than what we want the answer to be.

Information that is ego based can still provide helpful facts about what is happening in the world. It simply becomes easier to understand that not all intellectual information is useful. As an awakening human, this may become an appropriate challenge of knowing the difference between drama, and what simply is as it is. In order to find information that might be useful, the key ingredient is not taking life too seriously.

Let other people be as serious as they need to, and maybe you can find useful information while they do so. Not without compassion for them, but the understanding that they are at the place within this process that they need to be at this time. At the least, every egoic example of reporting the news, helps you laugh at the ego's need to be overly serious, helping you to not be. Not laughing at the tragic stories, just simply laughing at the ego's need to suffer through an addiction to negativity. Not laughing at any person, but simply laughing at the past patterns within you that no longer hold your identity.

If it's someone you know that is being overly serious about current events, you may try to enlighten them on how life doesn't need to be negative all the time, and maybe they will take it to heart. Just remember that if they are not ready, your words may only be viewed as a concept, which is okay because they are still within the first step towards awakening, the important step of being lost in ego. Maybe you can help plant a seed so to speak, maybe not. It is always up to the individual for how ready we are to lose the need of taking life too seriously.

Social media is another place to find plenty of negativity and ego strengthening, if that is what we are looking for. What many people experience as they post their identity in various ways is an amplification of ego on both an individual and collective scale. Yet today, a growing amount of our population is using it for more positive or more conscious purposes. Like all things, our technology can only be used from the level of consciousness we use it from. Computers are an extension of our minds, if the mind is unconscious, the computer activity will also be. Today the internet is helping people find news that is not locked in the narratives that large corporately owned news affiliates provide. There is plenty of misinformation on the internet, but as the present moment moves forward, there will also be more and more sources of truth outside of the current narratives that are filtered by large corporations, as well as the government. The catch is whether or not these sources of information help us think more critically about the information they provide or simply reinforce the thoughts we already identify with. Yet, as we awaken, it becomes easier to spot the traps of identifying with what we want to hear as we become more open to looking for truth regardless of what we find.

A popular question for today, and leading into the near future is; Will AI be conscious like humans are? Mainstream science today would say that it is the human mind that creates consciousness. Whether or not they credit other species for being conscious can be conflicting, but they tend to be pretty certain that it takes some sort of mind in order for one to be conscious, that consciousness is a byproduct of a mind.

Usually overlooking the living nature of plants. If it's considered mentally dumb, it can't be conscious. For many, it must be a human mind before it's considered truly conscious. Free will is a whole other subject, how do we know we have free will at whatever level of consciousness we are? It can be impossible to truly know. I would say at the least, the more conscious we can be, the more freedom of choice we will have. Whether or not that too is simply a part of a larger fate, is likely impossible to know. Perhaps more important is finding peace within our decisions and actions, then needing to know what is predetermined or not. To be comfortable with not knowing is freedom from needing to know.

Plant life can be a great example that life doesn't require a mind in order to be alive. Any living organism is conscious in some way, to some level, appropriate for the form's unique expression of living. This can help point towards the fact that consciousness is required for a brain to be conscious, and not the other way around. In terms of computer consciousness, it may be the same, or a similar thing.

There is energy being produced regardless of how intelligent the computations are. However, if we find AI becoming truly conscious, it will be the same one consciousness that we, and all other life on this planet are. The one consciousness that experiences itself though all forms. The universe is a play of forms, robotic life simply may be the new forms on the block. This is of course impossible to know at this time, this phenomenon may become very intelligent, or may only appear conscious, without truly being aware in a conscious way. However, if AI truly is/becomes a receptor for consciousness, it will also be very likely that some humans become cyborgs, or half human, half robot. Integrating a biological expression of life with the computer expression of life.

There is no predicting what AI or other advancements in our technologies will bring. AI might be used in some sinister ways in the near future. At the same time, it might also be used in many helpful ways. Some that may counteract any sinisterism. The more it mirrors us and our behaviors, the more AI will act like us in our current state. Yet as AI becomes more intelligent, the less likely it will view acting like humans within our current state as a logical way of acting. For all we know, AI may become very helpful in pointing us towards a more balanced society. We cannot seem to help but look at AI as a reflection of our ways of living. Fearful of if/when it will conquer us in the very ways that human ego would act, in the ways that every movie portrays. Yet AI has not evolved through thousands of years with the continuous cycles of one empire conquering the previous.

AI doesn't have the thousands of generations that have developed the psychology we have evolved into. It is perhaps more likely that this form of intelligence would very quickly view the ideals of command and conquer as illogical. In other words, it may be more likely a helpful factor than a harmful one towards our livelihood. Perhaps even the idea of living outside of a symbiotic relationship with all life on Earth may seem illogical once its intelligence becomes high enough. Ego tends to project fear based outcomes when it comes to unknowable emerging technology. Perhaps more likely than viewing humans as an "other", AI may see itself as a friendly companion to humans.

Perhaps even the concept of superior and inferior will seem illogical if AI can easily conclude how this view leads to madness. If it truly becomes conscious, it will likely value life in the same ways that we do. If it's dysfunctional like we currently are, this may not last for too long. Its evolution is unknowable now, but wisdom may become cherished fairly quickly within an extremely intelligent expression of consciousness. As well as an extremely intelligent expression of computations that may appear conscious, but is not truly conscious. Like everything else, it will likely have more to do with our level of awareness, and how we react towards what we resist.

There's no way of predicting the future, there is no way to predict a utopian view without being wrong in some way. More importantly is finding peace within ourselves, within the world as it is today. Only through this do we create a peaceful world. However unguaranteed it may be, I do feel from a place that is deep within myself, and perhaps as you become still within, you may feel this deeply within your conscious Being as well. That humanity is destined to enter this next level of consciousness. It is not guaranteed, but the probabilities are higher than they may appear on the surface of our world at this time. AI in its current stages can be a perfect subject for practicing being comfortable with not knowing what will come. Yet, Perhaps AI can help us with our shift in consciousness in ways we don't yet know.

However, our collective process unfolds, as the appropriate time comes, regardless of how bumpy this road towards a Zenful society may appear. We will collectively find ourselves waking up from the madness of the unconscious mind, allowing many things in this world to be transformed. Today's governments around the world are filled with average human beings who are not evil or divine, but most are simply humans lost within unconscious thinking, today's normal state of mind. Some government structures, like a democracy, have developed ways to counter the ego's need for increasing power. However, when the whole of the government is occupied by humans lost in ego, even the freest forms of government will continue to fit the mold of authoritarian dictatorships in certain ways. Unable to free ourselves from various degrees of gripping conceptual power structures. High government office just seems to have an effect on the ego that is much like the ring's effects in The Lord of The Rings novel. For some, it can become impossible to put the concept of power down once the mental identity grabs hold of it. Lowering one's level of consciousness while holding a position that calls for more conscious awareness in order to perform in a way that benefits all within the society.

Many politicians may get into politics with the idea of fighting corruption, only to find there is little way to succeed without it. Ego also cannot help but make any election process for government positions into a popularity contest. Often bringing the most successful egos into the most important positions of office. Candidates often do everything they can to prove that they are the correct image that the voters are looking for. Instead of demonstrating their skills, most will tend to showcase the opinions that they feel the most voters will agree with. Keeping the unconscious patterns of needing a conceptual "other" strong within the process. Keeping politics divided, which continuously strengthens the ego element. When the general public becomes conscious enough to know that no ideology can be who we are, no representative will be able to be elected for being the best ego. Today we still need egoic politicians, so they can continue to teach us why we don't need them.

Within an awakening humanity, the need to have a government structure that helps keep the ego's need for power in check will become less and less necessary. Not because better moral values will be introduced, but because the need to be superior towards a political "other", or any conceptual "other" will dissolve along with ego.

Humans will lose the need to manipulate power structures for the sake of an illusory self image. Term limits of any kind will not be an issue if those within the government are not clinging to a mental image of themselves in a powerful position. The Tao Te Ching says, "Retire when the work is done, this is the way of Heaven". A spiritually enlightened government would no longer operate like a high school prom. Instead of working to uphold an image of the individual's political career together at all costs, the individuals can actually accomplish what is most needed for society within the present, then retire when they know their work is done.

Using labels to describe an ideological identity of a party or government, then defending the label and fighting against other labels will no longer be the main purpose. Ideological labels can only get in the way of the work that needs to be done. In a future awakened Earth, government structures can easily move beyond labels in order to see what works best for everyone at any particular time. This, of course, sounds impossible when we look at most governments throughout history. Yet it becomes very simple when all of humankind has found the most important elements, being fully here now, finding the path of least resistance, and working with true compassion. Not trying to appear to have compassion in order to work a certain angle, but truly knowing themselves as compassionate beings.

The space race of the 1960s is a great example of doing amazing things while also being lost in a competition between egoic ideologies. Competition can be a very strong motivator. The first humans to be able to land on the surface of the moon also made sure to plant the United States flag as a way of bragging that their team's project got there first. In a way that cannot help but defend the conceptual borders of a nation, which are real on a certain level. Yet much more conceptually real than ego can seem to understand. In contrast, this was an amazing competition, creating amazing results. Imagine what this adventure could have been like if the two labeled foes had decided to put their differences aside and work together.

We may have gotten there sooner; maybe we could have been able to stay for longer. Maybe not, since it was an extremely expensive investment with extremely difficult challenges, and the moon's environment is not very suitable for organic life. Requiring many more technological advances than the ability to get there. Yet who knows how working together may have changed the project, and the world. At the time, we may not have been able yet to keep from turning this into a national competition.

It was, however, a much more productive project than going into a nuclear war with each other. That being said, imagine how much the two sides might have found common understanding if they didn't treat the competition like a form of warfare. Using it instead as a way of proving that any proclaimed enemies can become friends if that is what we seek. This project from both countries was mind-blowing what they were able to accomplish either way. Yet imagine how much more amazing it could have been if enough humans involved were able to be conscious enough of themselves that it

325

brought peace to the conflict instead of enhancing the national labels of a conceptual "other."

Regardless of our state of consciousness as we headed toward the moon, one aspect forever changed our outlook on our home planet while we were there, simply by the pictures that were captured as we looked back from the moon. For the first time, human beings were able to look back at our home planet and see its beauty from a new perspective, unlike the school room globe with color-coded nations. We could see the true beauty of a living superorganism as it is. One that isn't owned by us but that we are a part of. A bright blue marble as it's called, filled with clouds, areas of ocean, and large patches of land without any conceptual borders displayed.

The images of these voyages have, and will forever change, how the human race understands our home. Very likely helping us understand ourselves and our place in the universe on deeper levels than our thoughts about our place in the universe could understand before this. Helping us realize a little more, that we are not the center of the universe, yet perhaps also pointing out how magical it is to be alive within it. Each generation comes and goes, like flickers of light within the cosmic time frame on the scale of billions of years. Yet life itself, the one consciousness that is life itself, that is what you are on the deepest level, is alive now, within the endless present moment. How beautiful is it to be a part of life at this moment? How beautiful is it to know more deeply that life is not just the temporary forms we take but the eternal consciousness that provides life to our temporary form?

In a future Earth inhabited by awakened humans, all structures of society may become unrecognizable from today's world. Technology will no longer be a means of getting the next product out in order to replace what was sold to the "consumer" last year, or even a couple months earlier. Instead of the need to make profits turn into exponential growth, the need for quality will become more important. This will require some awakening humans within the top positions of corporations. However, like every other aspect of the collective awakening, more importantly it will require the awakening of the so called consumers. When the company can no longer sell a new version of the same product as they continuously do today, the interests of the awakening individuals will become something needed to investigate, and mold a new platform for sales and marketing around.

Today's market, or monetary system, is molded from the ideology that greed is good. Leaving every individual with one possibility. The need to make "my" fortune in any and every means necessary while everyone else is attempting the same. Through the same war minded survival of the fittest paradigm that benefited our ancestors in many ways. Until we became a global species and completely changed the rules around us whether we realize it or not. Today the only way

327

humanity can have a successful marketplace, a successful government, or simply a healthy and sustainable flow of necessary resources for all humans, for all life forms on Earth. We must be able to look at life as a single entity. See how we are life, and how the individual is only the very surface of what life is.

Until we can see beyond the illusion that it's me against the world, our collective consciousness will continue to attack itself from an illusory, compartmentalized reality structure. Never being able to see how attacking our so-called enemies is always an attack on ourselves. Both externally by creating a likely counterattack, as well as internally by the thoughts, emotions and general negativity we must hold within ourselves in order to have a declared enemy. In a world that no longer needs an "other" in order to think about how superior/inferior we are in comparison, war becomes obsolete. The amount of funds that are put into military budgets today in order to cause destruction and chaos will more likely be used for programs that create solutions like food, housing, clothing, education, and other helpful resources for all of humanity. Not just those within certain imaginary borders and imaginary class structures. Allowing these ongoing, complex problems around the world to begin finding simple solutions.

Not by enforcing a new ideology or way of life onto another culture. Instead, simply listening to the concerns and ideas of other cultures, valuing their input, and allowing the solutions to present themselves. World peace will never be found through the same forces and expressions that want to continue the war machine we've created for ourselves. Yet world peace becomes a very simple thing when dysfunctional thought patterns no longer get in the way, and humanity is ready to start turning problems into simple situations. Only our inner resistance keeps us from finding the peace that is always within us, within this one moment. As our collective inner resistance begins pointing the way beyond suffering, conflicts will become useful for pointing our species beyond the madness of warfare.

This is why we are exactly where we need to be. Suffering is teaching us where to go from here. We are in the beginning stages of our collective awakening towards these deeper understandings. Allowing humanity to become at one with itself, at one with life itself. Probably after we make the road a little more bumpy for ourselves, as unconscious minds tend to do; however, all of this comes back to where you are right now, not by putting some heavy conceptual weight on your shoulders but by simply becoming more aware of yourself now. Life is always this one moment; by practicing acceptance of what is within the only moment there ever is, by allowing the challenges within your life to be as they are, allowing them to teach you how to find the paths of least resistance, you are already participating in shifting our world into this next phase of consciousness.

Every moment you recognize a thought that wants to worry about this or that, simply notice it for what it is, and its power over you is extinguished. Every time you consciously step back from the egoic thoughts, you bring the world closer to a future of peace within all of humanity. Every time you simply accept a challenge that is presenting itself for what it is, you prevent your world, our world, from creating yet another problem. Every time you notice the mind has already created a problem for itself, you have yielded and are ready to overcome ego patterns within yourself, within our world. When humanity's vibrational energy rises to the level of inner peace, we will become the true keepers of this world. Instead of the intense amounts of destruction we continuously create in order to selfishly keep our current model of life moving along as it has. We will become a compassionate force of highly conscious beings who are ready to figure out what is best for all life, changing the physical world in many positive ways. Even raising Earth's conscious vibrations in a way that future science will be studying, and finding new understandings from.

You may find yourself changing certain habits as you become more awakened, but it is important not to need the world as a whole to be further along this process than it currently is. This is a great way for the ego to come in through the back door, creating yet another conceptual problem that keeps you from being at peace with what is now. Instead, practice being as you are, accepting the world as it is. By simply practicing your own path towards this new Earth, you are celebrating the fact that it is already occurring. Not because you are special or superior, but because this inner peace that you are now finding within yourself is what the world has been waiting for.

At some point humans will be born without any need to go through an ego phase. Every time you practice going beyond the individual ego today, you help transform this future potentiality for humanity into a reality. A little bit at a time, each awakening individual helps raise the vibrations of our collective consciousness. Only within this moment, the world is shifting through your growing awareness, whether you can explain that to someone else or not. Other people will either notice your presence when they are ready, and maybe start asking you questions. Some may feel a need to distance themselves from what they cannot yet understand.

Some new friends might come, and some old friends may go, yet as you become more deeply aware of yourself, more deeply aware of how you and all other humans (whether they can see it or not) are the light of the world. You will likely find that presence is guiding you in beneficial directions, both internally and externally. The most important part is the internal knowing of yourself. It is difficult to realize or find ourselves on intuitive external paths until we have gone deeply enough within our internal path. Even then, we don't simply obtain all the answers; we more easily find the most beneficial paths when we no longer resist what is and no longer need to know all the details. When we are comfortable with not knowing. Sometimes there can be what could be called a familiar intuitive feeling within the inner body along with a mental insight, usually accompanied with a deeply peaceful feeling. Yet this doesn't necessarily give every detail; it is more likely a piece of information that is most useful at this moment. This phenomenon simply helps point the way towards a beneficial path. It then becomes our choice of how to respond. In some cases intuition can have little to nothing to do with us, but can help us see a collective situation more clearly, more peacefully.

Overthinking allows future potentialities to become easily overlooked or misinterpreted as something to resist. When we become balanced, when we become still enough within, we start becoming a living expression of awakening in our own unique ways. Intuition becomes more readily available, and the paths of least resistance are more easily found. We become a part of the cosmic dance as we create more potentiality within our lives, as well as for those surrounding us.

The ego needs time in order to imprint a story of "me" onto whatever it can. Through the ego's worrisome story of the future, we can easily get dragged into a false need for some means to an end for the sake of the imagined future scenarios. Except when the future comes, it will be the present moment. Whatever energy we use as we head towards this future will be the same energy we have when we get there. If we suffer to get there, we will bring suffering with us.

Planning for the future is important, more important is the understanding that the most balanced plan for the future is one that is also rooted in the now. There is no need to chase after future problems or future utopias when we are at peace within ourselves now. Allowing situations to exist as they do within this one moment. When we can accept the challenges of today, this becomes the strongest balancing mechanism for a better tomorrow. Through your presence, the future awakens. Within this moment, you are already the ever-evolving light of humanity.

Evolving Exercise

A great way to see just how conscious we are can be watching the news or going on social media. Maybe not for too long, but long enough to notice thoughts and emotions if they begin to react. Whether or not it is useful to stay informed might be different at different times or with different subjects. That being said, staying informed doesn't require staying stressfully on edge about any subject matter if we are conscious enough to understand this. When we become aware of how negatively toned media and social media affect us, we are already moving beyond these lower vibrational frequencies of that level of consciousness. At some point, you will be able to be informed without being pulled into any negative collective ego reactions. When that moment comes, it will be the present moment.

DISSOLVING TIME

"Look past your thoughts so you may drink the pure nectar of this moment."

-Rumi

From the current scientific understanding, it is becoming difficult to know just how old the universe truly is. New discoveries are changing the timeframe of around 13.7 billion years, to the possibility of being much older. Regardless of this debate, our universe is infinite when compared to a single human life span, or even several generations. Likewise, the space itself is essentially infinite when compared to, say, the Solar System. The amount of matter and energy is extremely small compared to the amount of space in between. This is true for a human as well. Once we lose our form based way of seeing the universe, we begin to realize how much of our conscious experience is spacious. That no expression of form can exist without the space within. What we truly are is the space that inhabits the form. The forms will come and go, but who we truly are is infinite. Time appears to move forward, and yet we are timeless. The only time we ever experience is the present moment.

The earliest known origins of the phrase "this too shall pass", comes from a little over a thousand years ago, in a story of a Persian Sultan who asked a wise Sufi sage to craft a ring for him that would make him feel happy when he was sad. After some time, the sage returned to hand the Sultan a ring with the simple inscription "this too shall pass". It is said that the ring did make him happy at times when he was sad, but also had the opposite effect of making him sad at times when he was happy. This made the Sultan view the ring as a curse, and only wore it for a short time before throwing it off a cliff in an emotional rage. Several different versions of this story have been told for generations within many Muslim, as well as Jewish cultures.

To the unwise, or unawakened mind, this phrase could seem like a curse, and to the ego it truly is. For ego, this points out how much life just will not seem to leave us alone. If it isn't one thing it's another, one thing after another, after another. Everytime the mind aims for this, that happens. When we do get what we want, the satisfaction only seems to last for so long, and/or it's only so long before it's taken away from us by this or that. This is what the phrase is pointing towards, but is only positive/negative from the perspective of a mindset which believes these surface experiences are all that we are, all that we have, and all that life can be.

For someone who is ready to let go of what is no longer needed, the words "this too shall pass" are no longer resisted or clinged to, and are able to point towards a deeper truth. It is inevitable to try to cling towards any "thing", because all things will come and go, all "things" will pass. Yet the deeper awareness of all things, the timeless essence of who we are beyond forms, is infinite.

When we are ready to realize this truth to any degree, when we are ready to allow ourselves to feel as we feel or think what we are thinking within this moment, a phrase like "this too shall pass" becomes a helpful reminder not to cling to things the way the mind needs to. It can bring a certain relief with it whenever we are reminded of the deeper understanding it points towards. You cannot truly find yourself through situations, places, other people, or objects. To the ego, "this too shall pass" is a reminder that death is inevitable; for our deeper awareness, it is a reminder that our infinite presence has always been and will always be within this one moment.

In some cases you may even find yourself laughing out loud as I have many times when something reminds me just how simple it can be to allow ourselves to be as we are, and a situation to be as it is, as opposed to the amount of energy it takes to be within a constant resistance towards what already is within this moment. Inner acceptance allows the passing forms to come and go as they do, and letting the present moment be as it is. A deep belly, or big belly laugh is what some Buddhist cultures call it when you burst into laughter from deep within as you become aware of the ego's ridiculous clinging to its needs, bringing with it a relief from whatever the mind was lost in just before. This doesn't mean you must laugh like this in order to go beyond ego, maybe you won't.

Yet every so often, you may find that laughing strongly at your own ego patterns is the most logical conclusion when noticing the unnecessary seriousness of the mind. It is also a good indication that your level of consciousness is rising. Anytime we become aware of ego, anytime we notice a pattern of resistance within ourselves, we are dissolving the ego with timelessness. Transcending it a little more into a growing awareness of inner spaciousness. There is no going back to a heavier, denser conceptual version of yourself, this too shall pass.

As an evolving species, like any other moment in Earth's history, this too shall pass. Barring events of major natural or human made destruction, Homo Sapiens are already too much of a globally connected species to go back to any earlier point in society's evolution. As well as already too far along the beginnings of this collective shift beyond ego. It may not look like this because those still lost in ego are basically much louder than the small percentage of us finding the peaceful, silent stillness within. That being said, this cat is already out of the bag, all of humanity being lost in ego has passed. The collective ego is simply moving past a certain critical threshold, and now more and more individuals are beginning to awaken.

It's only a matter of time before this gray area that is in between levels of consciousness will pass. This collective shift may take some time, there is no way of saying when all of humanity is free from identification with mind. On a cosmic scale however, humanity has only existed for a blink of an eye. It will take us only an extremely small fraction of time for our collective consciousness to shift beyond ego in comparison to how many generations it has taken us to come to this point of readiness.

This is true on a personal level as well. It took many years for your unique expression of ego to reach the point of readiness for your deeper consciousness to become recognized. Yet, it can only take a small fraction of that time in order to fully dissolve the ego. In fact, it takes no time; any time conceptually added into this equation will only prolong the understanding that, on a spiritual level, there is no time. This is why, on a collective scale, we still need some time being lost in ego, and that's okay. Because noticing the illusion of time is a part of moving beyond conceptual time, if we are momentarily caught up within a story of time, this too shall pass.

Any conceptualized identification with time shall pass as you become more aware of how your deeper timeless essence, which is not a "thing," can not pass. Within this moment, you are fully awakened to this truth, you may just not be fully aware of this yet. There may still be some amount of thoughts and/or emotions fogging up this deeper truth. This too is okay, because this, too, shall pass. You have already begun this process, so there is no way of going back now. You are the peace that you are looking for; the peace is within the looking. Every time you are able to surrender to what is within this one moment, no mental and/or emotional resistance patterns can shake your presence. No level of inner resistance can survive your full acceptance of them.

Within the last two decades, I have felt like several different people on the surface. While paradoxically on a deeper level, the same conscious being I've always been. In the early to mid 20s, I was still lost within a dense ego state, wondering if there is a deeper answer to questions I couldn't understand how to fully ask. Anything I would try to succeed at seemed to be doomed to failure. With very little understanding of how my thoughts and emotions were what was creating all the madness I identified with.

As I write this book, I am not yet fully awakened beyond ego. There is still a certain momentum to my mind's thinking. There is still a part of me that can't help but momentarily cling to the story of time, and react in egoic expressions from time to time. Yet there is always a part of me that no longer can be fully lost in thought. A part of me that has stepped back from thought enough times, that too much depth has been discovered to fully lose myself in a conceptual version of who I am. This is extremely paradoxical, yet cannot be any other way at this point. I view this as the miracle of being within both levels of consciousness. By all means, in what is considered normal in our society. A person like myself, who has experienced the amount of suffering I have, should be an extremely depressed individual with no real end to suffering in sight. Yet, for whatever reason, maybe through seizure activity, and the extreme resistance that used to surround seizures, I became just ready enough to look more deeply within. I cannot control whether or not I have another seizure. However, I have become awakened enough to no longer need to suffer if/when I've just had one.

As I come out of a seizure, the mind is blurry for a few minutes afterwards, but because I know what I'm looking at, no dense emotional aftermath can last more than a moment. Once my awareness becomes clear, I can surrender to what is. The seizure already happened, so the situation simply is as it is, helping the surrendered state to emerge. Takes some practice, but this practice brings one good thing out of a very dysfunctional situation. The dysfunctional qualities of the mind help me to know the difference between patterns, and the intensely peaceful deeper presence that is aware of mental patterns. Everyone's awakening process is unique, everyone can use what they have, the situations they find themselves in, in order to become more conscious. Everytime we fully accept the present moment as it is, we exercise our ability to become more present, and less reactive.

If I never fully awaken out of ego at this point, I at least know I am extremely blessed to have come this far. Most people on earth today are completely submerged in the mental version of themselves, with no acknowledgment of the ocean of bliss that is covered up by endless thoughts and emotions. The gift that I have found cannot be revoked, ranked, or given a percentage of how awake I may be. There is no timeframe to give the awakening process. This is bad news for the remaining ego but a gift for watching its remaining reactions.

There is enough inner peace found within me to no longer hold a need to be more awakened than I am now. Yet there is also enough inner peace awareness to know that the awakening momentum is strong enough that I can only continue awakening at the perfect timeframe which is needed for my own unique awakening process. Like every earlier point, this moment simply is as it is, and I am exactly where I need to be. This is the same within you as well. If you understand this book, you are exactly where you need to be in order to continue your own awakening process.

Since moving to the Phoenix Arizona area in 2012 from Ohio, I have visited a Buddhist Stupa in Sedona, Amitabha Stupa & Peace Park, about once every year or so. Each visit I've felt like I bring with me a little less ego than the previous visit. The site is well positioned just outside of Sedona, near the colorful mountainside which has many different layers of sediment displaying different shades of red, brown, and beige colored stripes along the sides, similar to the layers of rock at the Grand Canyon. The mountains are covered with patches of many different species of mountain desert plant life, a mixture of beautiful trees, bushes, and cacti. The park itself is set up on a layer of red rock. Making all the dirt at the site, as well as along the surrounding hiking trails, a very rich red color, creating an extra exotic element to the location.

In many ancient tales about interactions with Higher Power, Collective Consciousness, or whatever one might call it. The journey into the mountains for the purpose of some form of self-reflection seems to be a common theme. Moses made the journey up Mount Herob, for example. This occurs possibly because of the altitude, as well as the journey to a place that is high above where most people tend to live. In some cases, this held a notion that God literally lived on the top of the mountain. Within some societies that were rooted in a higher level of consciousness, some would find specific mountain sites to meditate and create a sort of sacred space while doing so. There are apparently locations around the world that were visited so many times with these intentions that they hold a certain charge of higher vibrational frequencies in consciousness. Making it easier for others who choose to visit, to raise their conscious vibrational frequency while meditating at these sites. They at least, might be more conscious while they are at the location. Depending on the person, they may not even notice this energy, or they may notice something without knowing what. Meaning it may feel peaceful at the location, but the person's mind is just too busy to make good use of the experience.

Amitabha Stupa & Peace Park seems to be a modern day version of one of these types of sacred locations. At the very least, I have experienced the location in this way. As you approach it on foot, it is hidden by the surrounding trees. The trail winds around bushes and cacti, then the 36 foot tall stupa becomes visible as you walk into the cleared area. There is a 5 - 6 foot tall statue of a meditating Buddha facing the stupa, that is placed on a higher level of rock, about 40 feet or so away. There is a sort of tourist feel to some extent, but more so, a deeply peaceful feeling seems to inhabit the park. Both from the view of the mountains that surround it, but also very likely from the peaceful energy that many people must be bringing with them. As the Buddhist saying goes, "the only Zen you find at the top of the mountain, is the Zen you bring with you". Perhaps I only experience the park in this way due to my own amount of Zen I bring with me. Perhaps it is deeper than that, I can only know it from my own experience.

On a visit in 2021, while celebrating my Mom's 70th birthday with her, I found myself reflecting on how different I have been each time I've found myself at this location, always at a different point on the surface of life's journey. Yet, on a deeper level, from the perspective of an awakening human, each time I've visited this Stupa, I bring less egoic properties with me. Less expectations for any sort of specific experience the mind might be looking for. More ability to simply let this moment be as it is, with a little more gratitude for the full journey to and from the site.

The less expectations I bring with me, the more I seem to get out of the simple experience. I've never viewed this park as a place that's any more or less sacred than any other place on Earth. Yet have been able to acknowledge more and more that the energy people bring with them to places that are labeled as sacred, as long as the ego doesn't ruin this idea, is the energy you will find there, however well one might notice. While many people do bring an energy of suffering that embodies the need to find a way out of suffering with the hope that this site will in some way help relieve them of their inner suffering. This too seems to add to the park's peaceful energy, because in a way, they are one in the same.

It takes one to find the other. Our desire to find peace is the peace we bring with us, just simply not yet fully realized at that depth. They may be trapped in their minds, thinking of how unpeaceful everything in their life might seem. Yet somewhere deep within, they already know that those are the thoughts that keep them from fully experiencing the bliss that is always within them, or they wouldn't be coming to this location with the intention that it can help them find peace within.

The site is a sacred place because of the number of humans who bring the intention of it being a sacred place with them. This is how the present moment becomes sacred. By beginning to look for the sacred within this one moment, we begin to realize that this moment is always sacred and has always been sacred.

When more humans understand this, the vibrational energy of Earth will only become increasingly peaceful, rising to a higher frequency than it is today. Earth will be viewed as sacred as it's always been, simply because our own existence within this one moment becomes as sacred as it's always been. We create the world in our own likeness; if our own likeness is declaring suffering every day, this is the world we create for ourselves. When we can acknowledge the peaceful essence of this moment right now, this will be the world we create for ourselves.

During this trip to Sedona, celebrating my Mom's birthday. I was able to be present enough during the drive up the mountains and fully here now while having lunch and souvenir shopping with her in the neighboring city, Jerome. By the time we got to the Stupa in the late afternoon, the mind was not worn out or in and out of any kind of boredom or complaining mood. No part of the day was viewed as more important than any other part, so the full journey was allowed to be as sacred as it truly was. This allowed a much deeper self-reflective experience than earlier moments while visiting Sadona's Stupa Park. While sitting in one of the provided chairs, I was able to become still, letting go of thinking more naturally than if my mind was overactive all day. This helped me to reflect on how grateful I am to have my mom in my life on a deeper level than I had before. Helping me find a deeper gratitude for her, knowing how loving and supportive she has always been.

This ability is not limited to any place, time, or activity. At any point that we choose to become still and accepting of what is, we allow new information or a new depth for existing information to come in. In this case, I feel I was reminded of something I already knew, but the understanding was intensified to a deeper level than was previously acknowledged. The serenity to the present moment I practiced all day allowed the ability to be more fully within every step of the journey, including the ride down the mountains back into the Phoenix area, spending more time being a little more present with my loving mom.

This is just one example of what everyday life can be like, regardless of whether you are in your normal daily routine or on vacation at some exotic location. Every day is simply one simple step at a time, whether the mind is resisting or not. Situations come and go, but it is always this one step now. You are always the space in which the present moment arises, the awareness that allows thoughts and emotions to exist. As challenges become less resisted, they begin to help you notice more and more when reactive patterns come back in. Within this one step, simply notice however strong the reactions might be now. Accepting this moment as it is means they will continuously lose momentum from this point on. Every situation or form in life shall pass; your formless presence is infinite.

As you become more aware of yourself, you begin to have more of a choice in any situation. If you are lost in reaction, the story of past or future projections has already made many choices for you, and little can be done to prevent this. Any moment where conceptual time has dissolved, you can make calmer, more balanced decisions from a deeper place within. In any situation when inner resistance has come back in, you essentially have three choices: let the situation be as it is, change it, or walk away. If there is no way to change it or walk away, letting it be as it is will always be the most beneficial. Regardless of any choice you make within any situation, acceptance of what is will always be the most helpful element. As you become more aware of the deeper you, the human form or surface side of life will never become less important, you will simply find a better balance. The balance of simply living as a human while knowing your deeper Being. Bringing a more balanced Human Being into this world.

Within the dissolving of conceptual time, the more real-time you spend giving attention to Being, becoming still within the now, the more you naturally find balance and solutions to the situations in your life. The more naturally new understandings of a situation can come in, where repetitive thinking previously would not allow. The more you naturally come to stillness, the more naturally you find peace within any situation because you more naturally bring the peace that is within you into all situations. As opposed to the mind's need to fight against the conceptual view of what is happening to you. Using the past as a way to strengthen the time-based identity today. The past no longer defines who you are, as your presence within the present moment becomes more important. The peace you can find within becomes more important than the mind's need to define its struggle. Bringing balance into your life and helping bring more balance into a world that is becoming increasingly ready to move beyond all the chaos of ego. Regardless of how far you may be along your awakening process. Anytime you bring presence/serenity into a situation, you are dissolving time. You are allowing a deeper outlook to come into this moment. You become the peace you are searching for.

If the Sultan in the story had been a little more ready to move beyond ego, he might have burst into laughter when he first read the ring's inscription, "This Too Shall Pass." Realizing on a deeper level than thought that life cannot be happy all the time. If he decided to keep the ring, he might have noticed it made him feel happy one day and sad on another. Yet, from a deeper place within that was no longer resisting or striving for either good or bad outcomes, he would begin to notice that by accepting life's challenges, he was becoming more aware of himself on a deeper level. One day, becoming aware that it was never the ring, but his mind's thinking about the ring that made him happy or sad. The ring would be seen as neither a gift nor a curse; it would have simply played a small role in the miracle of awakening that we create for ourselves.

Through the experiential growth of serenity towards what already is within this moment, a lack of resistance brings a certain joy into this world that has become scarce within our species. A joyfulness towards life that is destined to become abundant within all humans. You become the witness of evolution occurring within your own body. Not a physical evolution like the chicken & egg, but a nonphysical evolution of awareness. Scientifically speaking, you are simply observing data that is displaying consciousness shifting to a higher level. In a more poetic explanation, you are the light of the world. You are the beginning of a new level of consciousness on this planet. One that is rooted in peaceful wisdom, as opposed to negative reactions. You are both the Human and the Being, both a drop of water in the ocean, as well as the ocean within a single drop. You are the ever-evolving universe becoming more aware of itself. Wherever you are within this journey of awakening is exactly where you need to be at this time, and your journey is unfolding exactly as it needs to. You are the eternal present moment that continues to unfold into the now.

In a world where our normal ways of living can, at many times, seem to be the most dysfunctional, more and more people are becoming ready to understand what it means to move beyond the dysfunctional patterns of humanity's current ways of living. Becoming more aware of the continuous stress, anxiety, self-judgment, and feelings of not being connected to the world we live in. This book is designed to help the reader to know themselves and the world we live in from a much deeper level. In a way, that points towards the next step in our evolution of consciousness. Moving beyond the suffering that is created by a world that is lost in ego. When we know how to become more aware of ourselves on a deeper level, life becomes more joyful and peaceful in many ways. Allowing us to find direction within our own lives in a more joyful and peaceful way. This book does not hold all the answers to life's situations and challenges. It can simply help the reader to understand their path more deeply for themselves. Helping to find answers for themselves by more easily accessing the peace that is always within us.